Force Majeure in the Hydropower Industry

Daniel Constantin Diaconu

Force Majeure in the Hydropower Industry

Concepts and Case Studies

 Springer

Daniel Constantin Diaconu
Faculty of Geography
University of Bucharest
Bucharest, Romania

ISBN 978-3-031-27401-5   ISBN 978-3-031-27402-2   (eBook)
https://doi.org/10.1007/978-3-031-27402-2

This Springer imprint is published by the registered company Springer Nature Switzerland AG
The registered company address is: Gewerbestrasse 11, 6330 Cham, Switzerland

This book is dedicated to my children Dragos Cristian and Diana Catrinel, for understanding and support, and with apologies for the little time spent together.

Preface

The energy market will be the vector of progress of all economies. The diversity of production sources, users and the geopolitical environment will lead to multiple contractual agreements between all parties involved. In this complex economic universe, unpredictable and sometimes disastrous phenomena of natural or anthropogenic origin are also present. Under these conditions and in certain situations, the fulfillment of the assumed contractual obligations will not be fulfilled. Force majeure is a legal way of managing risk in a contractual agreement between two parties. The unpredictable events were commonly mentioned used to be terrorist attacks, wars and epidemics, but in the last period of time, extreme natural phenomena have become the most invoked reasons to terminate a commercial contract.

This book is written based on personal experience as a technical expert involved in the process between a large energy producer in Romania and several beneficiary companies. The contents of this book intend to bring to highlight the complete contexts in which force majeure was invoked to terminate contracts of all kinds. I believe that this experience will be useful both to energy companies and to the beneficiaries or traders of electricity, and also to the courts that are put in charge of the decision-making in such cases.

There are many law firms or legal departments that can provide legislative advice to companies. But the challenge is to correlate the contractual specifications with the extreme cases that may occur and that may affect the performance or termination of the contract assumed by the parties.

I would like to thank all the lawyers I worked with throughout the process for their interest in the technical details and the judges who made the effort to understand these unique issues in the case law between companies.

Romania, Bucharest
2022

Daniel Constantin Diaconu

Acknowledgement The author thanks the publisher, Springer Nature, for providing the opportunity for this publication.

Contents

Chapter 1
The Energy Context in Romania

Abstract Water and energy are major components of the development system of human society. Given that the interdependence between the two components is very high, the demands on one element exert indirect pressures on the others as well. Global changes in climate, the structure of the economy, and consumers' habits are making the energy–energy system a permanent change. Knowledge of this context of its mechanisms must be the basis for future decisions.

Keywords Water · Energy · Sustainability

Water and energy are two vital components of the sustainability of increasing the number of inhabitants on Earth. Under these conditions, the exploitation of the resources we have must be carried out efficiently and sustainably to not jeopardize humankind's future development (Mikulčić et al. 2020; Ke et al. 2022; de Oliveira et al. 2022; Grubert and Marshall 2022).

Emerging economies will account for most of the new demand for water and energy, driven by the rapid growth of the economy and population, the need to generate more goods and services and to intensify policy toward electrification, water supply and sewerage networks. Only the engine systems used in the Chinese industry will account for almost a fifth of the increase in the total international electricity demand by 2040. A similar increase in space cooling needs is expected, with the number of home air conditioners in developing economies rising to 2.5 billion units, from about 600 million today (https://www.iea.org/reports/world-energy-outlook-2018/electricity) (Zhang et al. 2022). Accelerating the absorption of electric vehicles and electric water heaters could rapidly increase demand in developing economies.

In 2016, primary energy consumption in the European Union (EU) was 1543 million tons of oil equivalent, 4.0% above the 2020 target.

Between 2005 and 2016, primary energy consumption in the EU fell by 10%. This has been due to improved energy efficiency, increasing the share of renewable energy sources (hydro, wind and solar photovoltaic energy) in total energy consumption, economic recession and climate change (https://www.eea.europa.eu/data-and-maps/indicators/primary-energy-consumption-by-fuel-6/assessment-2/) (Chen et al. 2021; Krkošková 2021).

Based on preliminary estimates from the European Environment Agency, in 2017 primary energy consumption was 1563 million tons of oil equivalent in the EU. This represents an increase of 1.3% compared to 2016.

Fossil fuels (including non-renewable waste) continued to dominate primary energy consumption in the EU, but their share fell from 78% in 2005 to 72% in 2016. The share of renewable energy sources almost doubled in the same period, from 7% in 2005 to 14% in 2016, increasing at an average annual rate of 5.4% per year between 2005 and 2016. The share of nuclear energy in primary energy consumption decreased slightly, from 15% in 2005 to 14% in 2016 (https://www.eea.europa.eu/data-and-maps/indicators/primary-energy-consumption-by-fuel-6/assessment-2/).

The EU's dependence on imports of fossil fuels (gas, solid fuels, and oil) from third countries increased slightly, from 52% in 2005 to 54% in 2016, expressed as a percentage of total gross domestic energy consumption plus bunkers. In 2016, oil accounted for 59% of total net imports, gas 30% and solid fuels 11%.

Primary energy consumption in the EU increased by 9.2%, from 1570 million tons of oil equivalent (Mtoe) to 1713 Mtoe between 1990 and 2005. Between 2005 and 2016, primary energy consumption in the EU decreased by 10.8%, reaching 1583 Mtoe in 2016. Various factors have caused this decline, in particular improving energy efficiency, increasing the share of energy in hydro, wind and solar photovoltaic (PV) energy, the economic recession and climate change (http://www.transelectrica.ro/web/tel/productie) (Wen 2022).

In line with the Energy Efficiency Directive, the EU has set a target of limiting primary energy consumption to a maximum of 1483 Mtoe by 2020. In 2016, the EU's primary energy consumption was below the linear trajectory between 2005 and 2020 targets. However, the sum of all 2020 targets for primary energy consumption by the 28 Member States was 1526 Mtoe (https://www.anre.ro/ro/energie-electrica/rapoarte/puterea-instalata-in-capacitatiile-de-productie-energie-electrica). It is 43 Mtoe (3%) higher than the EU target for primary energy consumption in 2020 of 1483 Mtoe. Based on preliminary estimates by the European Environment Agency (EEA), in 2017 primary energy consumption in the EU was 1563 Mtoe, 1.3% higher than in 2016. This level is 34 Mtoe (2.2%) higher than the level set in the linear path toward the 2020 target (Dascalescu and Kleps 2010).

The national energy system consists of all interconnected electricity in producers and consumers. The data published by Transelectrica company in 2019 indicate a capacity of 24,406.01 MW of installed capacity. The higher installed power uses the hydropower potential of watercourses. The 943 hydropower groups have a total installed capacity of 6758.78 MW (Table 1.1).

The diversified production of electricity in Romania aims to ensure a supply of basic, permanent energy and peak energy, when higher amounts of energy are required for relatively short periods of time.

Thus, we meet operators that operate continuously, stopping and restarting production groups lasting over time and being very expensive as well as operators operating in periods of wind or sun. Hydroelectric power plants with significant water accumulations operate when needed, their start and stop lasting only a few minutes. In addition, they are the only ones that can provide services regarding the safety of the

Table 1.1 Situation of the installed capacity (MW) at the national power grid level from 01.04.2019

	Combustible	Groups	Installed power	Installed capacity	Net power	Permanent discounts	Power available
1	Coal	38	6232.20	4822.20	4128.00	1691.00	4541.20
2	Hydrocarb	254	5456.07	3565.98	3044.72	2078.73	3377.34
3	Water	943	6758.78	6696.82	6317.84	382.72	6377.34
4	Nuclear	2	1413.00	1413.00	1300.00	0.00	1413.00
5	Eolian	123	3031.57	3031.53	2977.19	24.85	3005.63
6	Biomass /biogas	53	131.98	131.15	122.32	4.01	127.95
7	Solar	633	1382.36	1382.36	1297.59	61.35	1320.15
8	Geothermal	1	0.05	0.00	0.00	0.05	0.00
	Total	2047	24,406.01	21,043.04	19,187.66	4242.71	20,162.61

Data source http://www.transelectrica.ro/web/tel/productie.

national energy system and the quality of the energy supplied (Dascalescu and Kleps 2010; Nastase et al. 2017; Zelenakova et al. 2018).

The installed capacity in the electricity production capacities in Romania is approximately 19,581,543 MW, at the level of 2021. This capacity is structured as follows: 33.9% hydro, 21.2% coal, 15.4% wind, 14.5% hydrocarbons, 7.2% nuclear, 7.1% solar, 0.5% biomass and others in lower proportions (biogas, waste, residual heat, etc.) (https://www.anre.ro/ro/energie-electrica/rapoarte/puterea-instal ata-in-capacitatiile-de-productie-energie-electrica).

Any unit of electricity production, which uses fossil fuels (oil, gas, coal), nuclear, wind, solar or hydropower, also uses, directly or indirectly, quantities of water, in different proportions (https://www.anre.ro/ro/energie-electrica/rapoarte/puterea-ins talata-in-capacitatiile-de-productie-energie-electrica) (Connors et al. 2002).

1.1 Conclusion

The evolution of Romania's energy system is given by the available energy sources and by the structure of producers and consumers, in the context of Romania's integration in the energy market of the European Union. The changes registered during the last decade, under the influence of the implementation of measures imposed by global climate change, the transition to electric mobility, the abandonment of the use of fossil fuel, widespread migration and armed conflicts have also produced changes in production costs and the amounts of energy generated.

In this context, there is an upward trend in electricity consumption, an increase in sales tariffs, but also the closure of production units that use fossil fuels, an increase in the photovoltaic and wind sector, but also plans to expand energy production to nuclear.

References

Chen J, Guo Y, Su H, Ma X, Zhang Z, Chang B (2021) Empirical analysis of energy consumption transfer in China's national economy from the perspective of production and demand. Environ Sci Pollut Res 28:19202–19221. https://doi.org/10.1007/s11356-020-11983-7

Connors S, Gheorghe A, Schenler W, Vladescu A (2002) Risk related topics in strategic electric sector assessment under sustainability conditions: a Romanian case study (SESAMS-RO). In: Bonano EJ, Camp AL, Majors MJ, Thompson RA (eds) Probabilistic safety assessment and management, vol I and II, Proceedings, pp 1525–1530

Dascalescu N, Kleps C (2010) Functional, environmental, ecological and socio-economic effects of the hydropower developments as main renewable resource in Romania. J Environ Protect Ecol 11(2):701–708

de Oliveira GC, Bertone E, Stewart RA (2022) Challenges, opportunities, and strategies for undertaking integrated precinct-scale energy–water system planning. Renew Sustain Energy Rev 161:112297. ISSN 1364-0321. https://doi.org/10.1016/j.rser.2022.112297

Grubert E, Marshall A (2022) Water for energy: characterizing co-evolving energy and water systems under twin climate and energy system nonstationarities. Wiley Interdiscip Rev Water 9(19). https://doi.org/10.1002/wat2.1576

http://www.transelectrica.ro/web/tel/productie

https://www.anre.ro/ro/energie-electrica/rapoarte/puterea-instalata-in-capacitatiile-de-productie-energie-electrica

https://www.eea.europa.eu/data-and-maps/indicators/primary-energy-consumption-by-fuel-6/assessment-2/

https://www.iea.org/reports/world-energy-outlook-2018/electricity

Ke J, Khanna N, Zhou N (2022) Analysis of water–energy nexus and trends in support of the sustainable development goals: a study using longitudinal water–energy use data. J Clean Prod 371:133448. https://doi.org/10.1016/j.jclepro.2022.133448

Krkoškova R (2021) Causality between energy consumption and economic growth in the V4 countries. Technol Econ Dev Econ 27(4):900–920. https://doi.org/10.3846/tede.2021.14863

Mikulčić H, Wang X, Duić N, Dewil R (2020) Environmental problems arising from the sustainable development of energy, water and environment system. J Environ Manage 259:109666. https://doi.org/10.1016/j.jenvman.2019.109666

Nastase G, Serban A, Nastase AF, Dragomir G, Brezeanu AI, Iordan NF (2017) Hydropower development in Romania. A review from its beginnings to the present. Renew Sustain Energy Rev 80:297–312. https://doi.org/10.1016/j.rser.2017.05.209

Wen XL (2022) Energy consumption monitoring model of green energy-saving building based on fuzzy neural network. Int J Glob Energy Issues 44(5–6):396–412

Zelenakova M, Fijko R, Diaconu DC, Remenakova I (2018) Environmental impact of small hydro power plant-a case study. Environments 5(1), Article Number 12. https://doi.org/10.3390/environments5010012

Zhang KL, Lu HW, Tian PP, Guan YL, Kang Y, He L, Fan X (2022) Analysis of the relationship between water and energy in China based on a multi-regional input-output method. J Environ Manage 309:114680. https://doi.org/10.1016/j.jenvman.2022.114680

Chapter 2
Water for Energy Versus Energy for Water

Abstract Water and energy are the facets of the same coin. The two are inseparable, ensuring the social and economic development of mankind. The numerical growth of the population as well as the development of the society has led to the increase of water and energy consumption. Both are needed to increase agricultural and industrial production and daily comfort. The analysis of the duality of water and energy is necessary to understand the implications on sustainable development, their tariffs and also the implications on the environment. Both the aspects regarding the water consumption required for each energy branch and the energy aspect for the treatment and purification of the water destined for consumption were analyzed. All this for understanding the relationship between water and energy and making the best choices related to the integrated management of water resources by decision makers.

Keywords Water · Energy · Usage

Water and energy resources represent an interdependence of enormous significance for global economic growth, the well-being of the population and the existence of life on earth.

Energy is required for a number of processes related to the collection and treatment of water, its transport to users and wastewater treatment.

In its report on global risks, the World Economic Forum calls on experts to present a ranking of potential global threats according to their likelihood and impact (https://www.weforum.org/reports). Thus, starting with 2016, the following problems are presented: a failure of mitigation and adaptation policies to global climate change, a sharp rise in energy prices and a water crisis.

Water has long been considered an inexhaustible resource. The current reality indicates that we have less and less water, especially due to the deterioration of its quality. The amount of renewable water resources available to each country varies greatly globally. However, many countries face a certain stress in allocating water resources, given the particularities of the hydrological regime, water quality or the large number of inhabitants. More than a billion people live in areas with significant water stress, a figure that is expected to increase at least threefold by 2025 (https://unesdoc.unesco.org/ark:/48223/pf0000225741) (WWAP 2012).

Water stress is defined as when the annual supply of drinking water from freshwater sources falls below 1700 m³ per person; water deficit is when the quantity is less than 1000 m³ per person and the absolute deficit when the quantities are below 500 m³ per person.

Over the next 25 years, water demand is projected to increase by almost 10% from 2014, while consumption will increase by more than 20% over the same period (https://iea.blob.core.windows.net/assets/e4a7e1a5-b6ed-4f36-911f-b0111e49aab9/WorldEnergyOutlook2016ExcerptWaterEnergyNexus.pdf). Regional patterns of water abstraction and consumption may vary within wide limits due to the country's economy structure.

As I mentioned before, intensive agriculture is already the largest consumer of water globally, accounting for about 70% of total freshwater abstractions worldwide and up to 85% in some developing countries. Agriculture also leads to the deterioration of freshwater quality, especially surface and groundwater, due to the excess of fertilizers and pesticides used to increase production and control pests.

The increase in living standards and the number of people living in urban areas makes water sampling represent 13% of the total volume of water taken for use in 2014, and it is estimated that this amount will increase to 17% in 2040 (https://iea.blob.core.windows.net/assets/e4a7e1a5-b6ed-4f36-911f-b0111e49aab9/WorldEnergyOutlook2016ExcerptWaterEnergyNexus.pdf). Three-fifths of the growth comes from India, Africa and other developing countries in Asia (excluding China).

This increase in water demand is taking place in a context where more than 650 million people, mainly in sub-Saharan Africa, do not have access to a source of drinking water and 2.4 billion people do not have access to sanitation services (https://iea.blob.core.windows.net/assets/e4a7e1a5-b6ed-4f36-911f-b0111e49aab9/WorldEnergyOutlook2016ExcerptWaterEnergyNexus.pdf; https://www.unicef.org/romania/stories/unicef-and-sustainable-development-goals-0). One of the United Nations Organization's goals of sustainable development is to ensure the availability and sustainable management of water and sanitation for all (but we refer to domestic wastewater collection services and their treatment). Pursuing this goal requires investment in raw water treatment plants and domestic wastewater treatment. All this implies a higher energy consumption, which supports their construction and operation. The industry and the production of electricity through hydropower plants generally keep the volume of water taken (about 10%) and consumed, as before. This is due to the modernization of industrial technological processes as well as a temporary cap in the construction of large hydropower plants.

Water has an important contribution to almost all forms of energy production. Large amounts of water are used in the use of fossil fuels, nuclear and biofuels (Table 2.1).

The total amount of water taken by the energy sector, in 2014, amounted to 398 billion m³ (Fig. 2.1). The differences are generated by the cooling technologies of power plants, which are adapted to the fuel used. For example, gas-fired boilers require less water for cooling than coal-fired boilers. Also, in the case of renewable sources, solar or wind energy uses water to clean the panels. Geothermal energy has

Table 2.1 Amount of water taken and consumed according to the type of energy sector (Spang et al. 2014)

No.	Fuel[a]	Capacity factor	Water consumption factor m^3/GJ		
			Estimate	Min	Max
1	Coal	0.85	0.722	0.505	1.157
2	Gas/oil	0.85	0.768	0.589	1.157
3	Nuclear	0.90	0.757	0.610	0.936
4	Biomass	0.68	0.581	0.505	1.015
5	Solar	0.32	0.852	0.778	0.904
6	Wind	0.20	0.006	0.001	0.027

[a]Steam turbine, cooling tower

Fig. 2.1 Water withdrawals by the energy sector, at the level of 2014 (The United Nations World Water Development Report 2014)

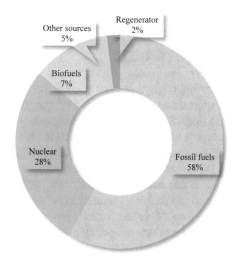

low water consumption if the quantities taken are reinjected after the heat exchange has taken place. Hydropower plants rely on water passing through turbines to generate electricity or store surplus energy produced. Most of the water taken returns to the rivers; however, water consumption in the hydropower sector varies depending on a number of factors, such as the type of technology (tank versus river), the size of the water tank, climatic characteristics and so on.

2.1   Water for Energy

Water is an important component for almost all forms of energy production. Hydropower are the main sources of electricity generation globally, providing 71% of

Table 2.2 Production capacity of hydropower (https://www.ren21.net/?gclid=Cj0KCQiAmeKQ BhDvARIsAHJ7mF4YNXNNPphbwYLxCaDPni1X6NBvrnOrBr73WpVK_oSNwr97H6cwoH MaAv1REALw_wcB)

Nr. Crt.	Country	TWh energy production	Hydropower potential GWh /year	Degree of use (%)
1	China	1126	2,140,000	41
2	USA	250	528,923	52
3	Brazil	382	817,600	48
4	Canada	376	1,180,737	32
5	India	120	660,000	21
6	Russia	160	1,670,000	10

renewable electricity. The installed production capacity reached 1064 GW worldwide in 2016 and generated 16.4% of the world's electricity produced from all sources. The highest energy production in 2015 was produced by China, followed by Brazil and Canada (Table 2.2).

There are many opportunities for hydropower development around the world; although there is no clear consensus, estimates indicate the availability of approximately 10,000 TWh/year of unused hydropower potential worldwide (https://www.waterpowermagazine.com/news/newsihas-2015-hydropower-status-report-available-for-download-4663043).

In Romania, the construction of new dams to include hydropower with significant electricity production capacities stagnated after the 1990. The lack of a strategic vision in this field, the lack of funds for the construction of such hydropower facilities and the problems related to the environment and the multitude of landowners made it impossible to start new projects. The only area addressed was that of micro hydropower, where following policies to subsidize renewable energy sources led to the construction of many such production units. In practice, it has been observed that for Romania, micro hydropower has a major negative impact on the environment as well as an insignificant energy production.

In the case of energy produced by biofuels, water is needed for the crops of plants used (rapeseed, soybeans, sunflower, castor, etc.) to produce primary energy. The scale of water use for biofuels depends on the water supply of crops (precipitation or irrigated).

It is estimated that approximately 2% of the total amount of irrigation water is used for the production of biofuels (Zarfl et al. 2015; WWAP 2009). However, there are significant opportunities to improve efficiency and reduce water demand. For example, the provision of subsidies to the agricultural sector often has the unintended consequence of encouraging farmers to use water inefficiently because of its abundance, thus depleting water reserves, often those underground (https://www.unwater.org/publications/managing-water-uncertainty-risk/) (WWAP 2012).

In the case of thermal power plants that use coal as fuel, the related water consumption is closely linked to the quality of the coal and its tailings. The need to be washed before use (in floating stations) and moistened for compaction and combustion leads to significant water consumption.

The same situation is encountered in oil exploitation. Water needs for conventional oil production depend on the technology used, geological conditions and the degree of secondary recovery. Injecting water as a means of increasing the recovery of oil from the ground can require significant volumes of water, even ten times more than the primary recovery, depending on the technology.

The amount of water required to extract natural gas is much lower compared to other fossil fuels.

Unconventional oil and gas production, which requires hydraulic fracturing, such as oil and shale gas, is not necessarily a large consumer of water, but can cause pollution of deep groundwater layers.

2.1.1 Electricity Generated by Nuclear Units

A nuclear power is a complex installation, equipped with one or more reactors, to produce electricity from thermal energy, obtained in turn by maintaining a controlled fission nuclear reaction, a process performed by the nuclear reactor. In a reactor, the heat produced by the cleavage of uranium atoms in nuclear fuel is taken up by heavy water (coolant) and transferred to light water (normal water), which is converted into steam in steam generators. Thus, the steam rotates the blades of a turbine that drives the electricity-generating generator.

Heavy water contains a much larger proportion than normal the deuterium isotope of hydrogen in the form of D2O ($2H_2O$) or HDO ($^1H^2HO$). The normal ratio of deuterium to hydrogen in standard VSMOW water is about 156 ppm. The HDO variety is also known as semi-heavy water.

In terms of macroscopic and chemical properties, heavy water behaves similarly to normal, or "light" water, but constituent hydrogen atoms contain an extra neutron in the nucleus because deuterium, or heavy hydrogen, is an isotope of hydrogen. As a result, its density is about 11% higher. Pure heavy water does not show radioactivity. The properties of heavy water are different from those of ordinary water, with a boiling point above 100 °C (101.4 °C), a freezing point below 0° C ($-$ 3.8 °C) and a maximum density of 1.2 g/cm^3) at $+$ 11.6 °C. It can be obtained in the laboratory by hydrolysis and distillation.

The source of water that ensured the production of this type of water used in the nuclear power plant facilities was from the Danube. It was filtered and treated and reacted with hydrogen sulfide in a 2-stage production line. The bitherm process (stage I and stage II), at two different temperatures, is concentrating on the first step from 0.0143 to 0.14%, and in the second step from 0.14 to 4–7% heavy water. The final concentration occurs in vacuum distillation, where heavy water for nuclear use is obtained.

The reactors used at the Cernavoda nuclear power plant, Romania, use about 10 tons of heavy water annually/reactor, and for their filling and commissioning, about

550 tons of heavy water/reactor are needed. The CANDU-type reactor (is a nuclear reactor of the PHWR type—Pressurized Heavy Water Reactor), i.e., a reactor in which heavy water, under pressure, fulfills the dual function of cooling the set of bars that form the nuclear fuel and moderating the neutrons resulting from spontaneous fission of uranium used as fuel. The CANDU reactor has a more efficient efficiency than the other types of reactors that use ordinary water, because it consumes about 15% less uranium. The average price of a kilogram of heavy water is about 500 USD.

The two functional reactors from Cernavodă have a total installed capacity of 1413 MW, generating electricity with a reduced flexibility in resizing/correcting production if at the other end of the technological flow there are no loyal consumers and with a constant operation. In terms of costs, it is the second cheapest producer of electricity.

2.1.2 Water Consumption in Coal-Fired Power

In the United States, in order to produce and burn the 1 billion tons of coal it uses in energy production, the mining and utility industries extract 55 billion to 75 billion gallons of water annually, according to the Geological Survey of the United States of America (Lance 2004; Crane-Murdoch 2010).

According to the United States Geological Survey (USGS) data, water extraction from thermoelectric power sources accounts for 49% of total extraction in the United States in 2005, which is 201 billion per day (Crane-Murdoch 2010, 200B Gallons of Water Drawn Each Day for U.S 2011).

The 2012 report "Burning Our Rivers: The Water Footprint of Electricity" estimates that one mega-watt hour (MWh) of coal-fired electricity extracts approximately 16,052 gallons and consumes approximately 692 gallons of water (Crane-Murdoch 2010; Wendy et al. 2012; NETL 2010).

However, these consumptions are only the tip of the iceberg, in terms of the impact on the water resource of coal-fired power plants. Sterile coal dust often reaches watercourses, causing great damage to biodiversity. Coal combustion produces most of the carbon dioxide and other greenhouse gases, which accelerates global climate change and reduces the world's freshwater reserves. And water is used daily in mining operations to cool and lubricate mining machines, wash transport roads and truck wheels to suppress underground coal dust that might otherwise ignite.

Water is used by thermoelectric generation (coal, natural gas and nuclear) to produce electricity by transforming water into high-pressure steam to drive turbines. Once this cycle is completed, the steam is cooled and condensed back into the water. Some technologies use water to cool the steam, increasing the water consumption of an installation. Because cooling systems do not recycle cooling water, this generally leads to very large volumes of daily water uptake. According to the Energy Information Administration, there were 719 single source and single use water cooling systems, 819 recirculating cooling systems and only 61 dry (without water use) and hybrid cooling systems installed in the US in 2012 (NETL 2010; Averyt et al. 2011; NDRC 2014).

In the production of coal, water is also used for cleaning and processing the fuel itself (coal float-washing stations). Researchers at Sandia National Laboratories made the estimate higher, finding that the typical 500-megawatt coal-fired utility burns 250 tons of coal per hour, using 12 million gallons of water per hour—300 million gallons per day—for cooling (Crane-Murdoch 2010).

2.1.3 Water Impacts of Modern Bioenergy

FAO (24) defines biomass as material of organic origin, in non-fossilized form, such as agricultural crops and forest products, agricultural and forestry waste and by-products, manure, microbial material and organic industrial and household waste. Biomass is used for food or feed (e.g., wheat, corn, sugar), materials (e.g., cotton, wood, paper) or for bioenergy (e.g., corn, sugar).

Currently, energy from biomass accounts less than 2% of US energy production; however, policies to promote renewable energy could lead to increases in biomass energy production. The major hidden costs of biomass energy production are short-term carbon emissions and related water consumption.

The Electricity Research Institute estimates the water consumption required for cooling at biomass plants at 480 gallons per MWh for plants with wet cooling towers (NDRC 2014). This would average 183 million gallons per year for a 50 MW plant operating at 87% capacity.

The highest water consumption is the necessary for the development of biomass. Accurate estimation of water volumes is quite difficult given that much of the water absorbed by plants is then released as vapors into the atmosphere (Table 2.3). In some countries, the need for crop water is completely or almost completely covered by irrigation water. These crops are sugar cane from Argentina (96%) and Egypt (92%); wheat from Argentina (100%), Kazakhstan (98%) and Uzbekistan (98%); potatoes and barley from Kazakhstan (100%); sorghum from Yemen (100%); and soy from Brazil (95%) (Gerbens-Leenes et al. 2009).

In general, the impact on water, generated by the production of electricity based on biomass, remains similar to fossil fuels, with high water consumption, ranging from 0 to 1800 L/MWh (Fingerman et al. 2011). Water requirements for biofuel processing continue to decline with the improvement of the technology used. The use of water per ton of raw material has dropped dramatically for both maize and cane sugar ethanol. For example, in the case of ethanol-sugar factories in southeastern Brazil it fell from 15 m^3/Mg of bagasse (bagasse = the dry pulpy residue left after the extraction of juice from sugar cane, used as fuel for electricity generators, etc.) of cane before 2008 to < 3 m^3/Mg in 2008 (Martinelli et al. 2013).

While untreated effluents can create water quality problems, process water offers an opportunity for nutrient recovery and recycling. Installations for zero-discharge biofuels have been operating in the United States since 2006 and continue to expand worldwide. Technological improvements in water recovery and recycling

Table 2.3 Total weighted—global average water footprint (WF) for 10 crops providing electricity (m^3/GJ) (Gerbens-Leenes et al. 2009)

No.	Type of crop	Water consumption m^3 per GJ electricity		
		Total WF	Blue WF	Green WF
1	Sugar beet	46	27	19
2	Maize	50	20	30
3	Sugar cane	50	27	23
4	Barley	70	39	31
5	Rye	77	36	42
6	Paddy rice	85	31	54
7	Wheat	93	54	39
8	Potato	105	47	58
9	Soybean	173	95	78
10	Rapeseed	383	229	154

have progressed to the point where some facilities are able to use municipal wastewater and others have closed-loop water recycling (Food and Agriculture Organization 2006; Li et al. 2021; EEA 2019).

2.1.4 Photovoltaic Installations (Solar Plants)

Whether they are solar power concentrators (CSPs) or photovoltaic (PV) systems, solar power plants provide pollution-free electricity production with an impact on local water sources, which are comparable and often smaller than traditional energy generation fossil fuel base.

The water use requirements for solar power plants depend on the technology and climatic conditions on the site. In general, all solar technologies use a modest amount of water (approximately 20 gallons per megawatt hour, or gallons/MWh) to clean solar collection and reflection surfaces, such as mirrors, heliostats and photovoltaic panels (PV). For comparison, Solar Energy Industries Association says that, a typical family uses about 20,000 gallons of water each year, more than the amount of water needed per MW of photovoltaic generation capacity (https://www.seia.org/initiatives/water-use-management).

In all thermal power plants, whether they are on fossil fuels, nuclear or solar concentrators, heat is used to turn water into steam, which drives a steam turbine to generate electricity. The steam discharged from the generator must be cooled before being reheated and turned back into steam. The heat is dissipated from the boiler by evaporation, most often through a cooling tower. Wet cooling is the most common cooling method for power plants, as it is the most efficient and cheapest method of cooling available at the moment (https://www.seia.org/initiatives/water-use-management). Water-cooled parabolic solar power plants and solar power plants in the power tower consume about the same amount of water as a coal-fired or nuclear power plant (500–800 gallons/MWh) (https://www.seia.org/initiatives/water-use-management).

For example, the Nevada Solar One satellite plant consumes 850 gallons of water per MWh on a 360-acre site near Las Vegas, or about 300,000 gallons per acre per year (https://www.seia.org/initiatives/water-use-management). In comparison, Nevada's agriculture requires nearly 1.2 million gallons of water per acre per year—nearly four times the consumption of the solar power plant (https://www.seia.org/initiatives/water-use-management) (Maina et al. 2022).

Since 2015, China is the country with the highest photovoltaic energy capacity. In 2017, China surpassed the European Union in terms of total installed photovoltaic energy capacity. With over 44 GW newly installed of the connected photovoltaic system it reached a total power capacity of 175 GW or 34% of the solar photovoltaic electric power of 518 GW installed worldwide at the end of 2018. The European Union follows with a cumulative installed photovoltaic panel power of 117 GW or 23% of global capacity. This is down from 66% in 2012, when the cumulative installed solar electric power had just reached 100 GW worldwide (Jäger-Waldau and PV Status Report 2019). The dynamics of this field is very strong, being restricted in certain situations only by the energy transport infrastructure at regional level.

2.2 Energy for Water

The connection between water and energy is two-way. Not only energy production is the sector that needs intrinsically water, but also water supply needs energy. The supply of fresh water from surface and groundwater sources or its production through desalination, its transport and distribution to consumers, as well as the process of collecting and treating wastewater requires energy. The amount of electricity consumed for each component of the consumers' water supply process may vary quantitatively depending on the type of water source, water quality and the technical characteristics of the distribution network, wastewater collection and treatment.

The actual amounts of energy consumed to ensure the operation of the water supply system can be made quite accurately. However, forecasting the evolution of the amount of energy consumed in the future presents a series of variables that can lead to an increase or decrease in value. The depletion of groundwater reserves and a decrease in groundwater levels will increase energy consumption for extraction. Deteriorating water quality will require treatment processes that involve higher energy consumption. Increasing the height of residential buildings will consume more and more energy to be supplied with water. But, increasing the performance of pumps, new technological systems of treatment and purification, could lead to a maintenance or decrease of energy consumption used in these processes.

Currently, the electricity consumption for underground water extraction amounts to 0.2–0.8 kWh/m^3, depending on the depth of groundwater, maximum 0.02 kWh/m^3

for its treatment, 5–6 kWh/m^3 for the treatment of seawater by reverse osmosis processes and 0.2–0.7 kWh/m^3 for the distribution of water in the network to consumers.

Overall, approximately 120 million tons of oil equivalent (Mtoe) of energy were used worldwide in the water sector in 2014, almost Australia's entire amount of energy. About 60% of this amount of energy is in the form of electricity, which corresponds to a global demand of about 820 terawatt-hours (TWh) (or 4% of total electricity consumption). It is almost equivalent to the current electricity consumption in 2009 in Russia. The rest is thermal energy, half of which is used for pumps driven by diesel engines, mainly for pumping groundwater for agricultural purposes. The rest is used for desalination, mainly in the form of natural gas, almost exclusively in the Middle East and North Africa. Globally, it is estimated that water extraction will consume over 310 TWh of electricity per year and about 0.5 million barrels per day of diesel (https://iea.blob.core.windows.net/assets/e4a7e1a5-b6ed-4f36-911f-b0111e49aab9/WorldEnergyOutlook2016ExcerptWaterEnergyNexus.pdf). Almost half of the global electricity for water extraction is consumed in Asia, as it is the continent with the largest amount of water used. India is by far the largest user of water in the world, in part due to inefficient irrigation of agricultural land, and accounts for about a quarter of global water abstractions, although per capita use is well below that of the United States. Other developing countries in Asia are also large consumers of water, especially Pakistan, Indonesia, Thailand and Vietnam. Some countries, such as India and/or others in the Middle East, rely heavily on groundwater, which is reflected in their relatively high energy consumption (https://iea.blob.core.windows.net/assets/e4a7e1a5-b6ed-4f36-911f-b0111e49a ab9/WorldEnergyOutlook2016ExcerptWaterEnergyNexus.pdf). On the other hand, Europe, China and the United States meet the water demand of uses mainly from surface sources. Non-traditional water sources are below 1% worldwide, with technologies and high energy consumption mainly making saltwater resources exploitation inaccessible. Also, in the category of non-traditional sources is represented by wastewater which through proper treatment can be transformed into water usable in agriculture or even for feeding the population (Luck et al. 2015; Bijl et al. 2016; Wada et al. 2016; EIA (U.S. Energy Information Administration) 2012).

A "source" of water for the future will even be the water losses of the supply systems of the uses. These losses represent directly volumes of water and indirectly amounts of energy wasted. Smart cities will have distribution networks assisted by leakage monitoring systems to observe damage immediately, and water flows can be redirected to consumers so that damage can be remedied.

In value, these losses amount, in 2014, to about 34 billion m^3 of water in India, Asia (excluding China) 20 billion m^3, China about 19 billion m^3, European Union countries about 13 billion m^3, Africa 12 billion m^3 and the USA with about 8 billion m^3. Obviously, the amounts of energy are also consistent with the volumes of water. Thus, India can save an estimated amount of energy at 35 TWh, Africa about 14 TWh, China 10 TWh, the European Union 7 TWh and the USA about 4 TWh.

Wastewater treatment, globally, consumes about 200 TWh or 1% of total energy consumed. The complexity of substances discharged by users in domestic waters requires a much more elaborate technological process that consumes much more energy than one of the years 1970–1980. High energy consumption can also be generated because not all the amount of domestic water is collected and treated at present. It is estimated that today, on average, over 35% of municipal wastewater is not collected, a value that is much higher in developing countries, 60–95%. In addition, wastewater is partially treated, often only a mechanical or at most mechanical-chemical step.

The level of domestic wastewater treatment varies enormously around the world: while the primary stage is the dominant process in some countries from Asia and Africa, the secondary (biological) stage is now standard in OECD countries, many of which also benefit from the tertiary stage. For example, the Water Framework Directive issued by the European Union (EU) aims to increase the level of wastewater treatment. About half of the energy used in advanced wastewater treatment is consumed in the secondary stage, mainly to meet the requirement for biological stage aeration (Fig. 2.2). Other important uses of energy include pumping wastewater throughout the plant, as well as for sludge treatment. In general, energy consumption in the activated sludge treatment stage is largely recovered by energy cogeneration (biogas production). It is estimated that the current global electricity production from sewage sludge is about 6 TWh and thus covers about 4% of the world's electricity demand in the municipal wastewater sector. There are situations in which the energy produced by the treatment plant, through the generated biogas, can cover the entire energy consumption necessary for the treatment of tributary wastewater. The tertiary stage is usually a less significant energy consumer, but water quality standards that are becoming much stricter in developed countries have already led to higher energy consumption for this stage.

Countries with small amounts of water, especially North Africa and the Middle East, have sought to supplement their resources by using salt water, desalination and re-use. The simplest desalination is the process of separating saline water (sea water or salt water) into fresh water and concentrated salt. Desalination of salt water, given its low salt concentration, consumes only about a tenth of the electricity needed to

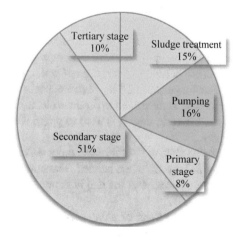

Fig. 2.2 Typical energy consumption in a wastewater treatment plant. *Source* (EIA (U.S. Energy Information Administration) 2012; EPRI 2002; Plappally and Lienhard 2012)

Tertiary stage 10%
Sludge treatment 15%
Pumping 16%
Secondary stage 51%
Primary stage 8%

desalinate seawater. Reverse osmosis is the technology that uses electric pumps to push water through a membrane to retain salt and obtain fresh water. At present, the reverse osmosis process is the most frequently installed technology, being stimulated by technological improvement and relatively low energy consumption. In 2015, over 65% of the global water desalination capacity was represented by reverse osmosis installations. In 2015, there were approximately 19,000 desalination plants worldwide, with a production capacity of approximately 15 billion m^3. There is almost half the capacity of desalination plants worldwide in the Middle East, followed by the European Union with 13%, the United States with 9% and North Africa with 8% (GWI (Global Water Intelligence) 2016).

The water reuse procedures are applied in the case of cities that do not have water resources but which in the centralized system do not deliver to the consumers drinking water, but only a water that can be used at home, but without being drinkable. Reused water is generally partially treated water that can be used in agriculture or reused in industrial technology cycles.

Electricity consumption in the water sector will increase by 2.3% per year in the future, being forecast to reach a total of 1470 TWh in 2040, the equivalent of almost twice the electricity consumption in the Middle East today. The largest increase in energy consumption is expected to come from the desalination process, as freshwater production from saltwater increases almost ninefold.

2.2.1 Pumped-Storage Power Plants

Pumped-storage power plants are reversible hydroelectric installations in which water is pumped upward into a storage tank. The force of the flowing water is then harnessed to produce electricity in the same way as conventional hydropower plants. Their ability to store electricity makes them an effective tool for overcoming the intermittent nature of wind and solar energy. They were designed in the 1970s to meet short-term requirements that thermal power plants or nuclear power plants could not provide by generally supplying basic energy to the system. Today, due to the imbalances generated by wind or solar energy suppliers, mainly, this balancing method has regained its importance in the storage-generation of energy. Pumped-storage power plants are based on two bodies of water, an upper tank and a lower one.

During periods of very high electricity consumption on the grid, the water in the upper tank is machined to produce electricity. At times when we have excess energy in the network, then it is used to pump water from the lower tank to the upper one (Fig. 2.3).

In this way, the water in the upper tank is a stock of potential gravitational energy, ready to be used when needed. The amount of energy stored is proportional to the volume of water and the height from which it falls (Pitorac et al. 2020; Rogner and Troja 2018).

The new model of using plants in combination with renewable energy has led to a revival of technology in China, the United States, Europe (especially Germany

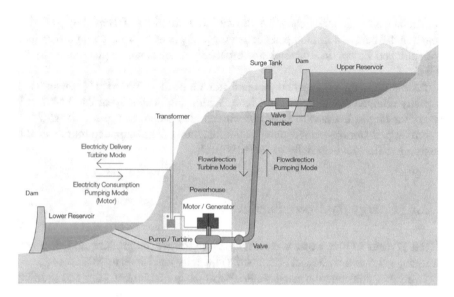

Fig. 2.3 Principle of operation of the system

and Norway), the United Arab Emirates, Morocco and other countries. In 2000, there were about 30 such power plants with a capacity of over 1000 MW world-wide. There are currently 170 units in Europe, but generally with a few hundred megawatts reduced capacity. With more than 100 projects underway, the International Hydropower Association estimates that pumped hydropower storage capacity is expected to increase by almost 50%—to about 240 GW by 2030. Despite the current development of specific batteries for the storage of electricity, hydropower tanks represent 95% of the total electricity storage worldwide and are also called "water battery" (Pitorac et al. 2020). It is a mature and reliable technology capable of storing energy for daily or weekly cycles and up to months, as well as seasonal applications, depending on the size and configurations of the project (Rogner and Troja 2018).

In essence, we can say that for the turbine to produce a power of 1 MWh, an average of about 1.25 MWh must be consumed to pump water to the upper tank. The result of this simple solution is a very high circus efficiency of 75–80%, which is a performance compared to other storage technologies.

The operating cost of a hydropower plant with pumped storage is low, compared to other types of power plants and has a long life of about 80–100 years. Such a hydropower plant of about 1000 MW requires approx. 100 permanent jobs, for operation and maintenance activities, after commissioning. Compared to a nuclear reactor within the Romanian company Nuclearelectrica, with 1400 MW of installed power, on a single platform (Cernavoda) it has 1950 employees.

The major disadvantage is the dependence of the design of the hydropower arrangement on the geomorphology of the site. For the most part, geological constraints are the main cause of difficult constructions. Given the geological constraints, there is a limited perspective for projects of hydroelectric power plants with pumped storage in Romania, and especially in areas with a predominantly flat relief.

At this time, the Guangzhou pumped storage plant is the first high head, large capacity pumped storage plant in China. With a total capacity of 2400 MW, it is one of the largest in the world (https://voith.com/corp-en/industry-solutions/hydrop ower/pumped-storage-plants/guangzhou-china.html). Voith supplied four reversible pumping units with a capacity of 300 MW each.

2.2.2 Energy for Water Purification

Seven percent of the world's electricity is consumed to produce and distribute drinking water and wastewater treatment (Plappally and Lienhard 2012).

Energy is consumed at each water supply stage, treatment, use, and purification cycle. The intensity of energy consumption (kWh/m^3) depends on the specific technological level applied in each stage of this circuit.

The activity of extracting raw water from natural sources (aquifers, rivers, lakes and oceans), treating it and bringing it into the parameters necessary for its use (drinking, industrial, agriculture), as well as for distribution to consumers and collection and wastewater treatment requires significant amounts of electricity in certain situation. Extracting water from underground aquifers requires primarily energy for pumping. Electricity (kW/h) is consumed when a unit volume (m^3) of water passes through a pump during its operation. The examples above would translate to a value close to specific ground water pumping energy use value of 0.004 kWh/m^3 per Km (https://iea.blob.core.windows.net/assets/e4a7e1a5-b6ed-4f36-911f-b0111e49a ab9/WorldEnergyOutlook2016ExcerptWaterEnergyNexus.pdf) (World Economic Outlook Database 2019; Cohen et al. 2004).

In the case of transporting water on open channels from the source to the treatment plant, the energy consumption is somewhere at 0.004–0.005 kWh/m^3 per Km (Plappally and Lienhard 2012; Dale 2004; Stokes and Horvath 2006).

The process of treating raw water and drinking it differs greatly depending on the quality of the water and the technology available. Energy consumption for drinking water from an aquifer requires, for example, a smaller amount than for desalination of salt water (distillation, reverse osmosis, etc.) (Wada et al. 2016b). Thus, we encounter a wider range of values of energy consumption compared to the unit volume of water supplied, depending on the water source or its initial quality. In general, the consumption is between 0.15 and 1.50 kWh/m^3 (Plappally and Lienhard 2012; Cheng 2002; Kneppers et al. 2009).

The vertical development of cities has led to an increase in energy consumption for pumping water to consumers. Starting with 1931, Empire State Building reached a height of 381 m, with 102 floors. In 2021, there are 75 buildings in the world that exceed 350 m in height, the tallest reaching 828 m, 163 floors, Burj Khalifa. The threshold of 1000 m (Saudi Arabia—Jeddah Tower) is expected to be reached shortly. Booster pumps performing the same process in California consumed energy in the range 0.015–0.41 kWh/m^3. This energy consumption range is very low compared to the energy ranges for Switzerland 0.25–2.5 kWh/m^3. Municipal water pumping in Ontario, Canada, consumed electrical energy in the range of 0.68–1.1 kWh/m^3. With a very high population density, China produced and supplied water in 1997 with an average energy consumption of 0.079 kWh/m^3 while this energy figure rose to 0.093 kWh/m^3 in 2004, simultaneously with urban development (Kahrl and Roland-Holst 2008).

Wastewater, residential, commercial and industrial, is polluted with a wide variety of pollutants. Domestic wastewater is treated before disposal or reuse. In the USA, the primary stage of wastewater treatment consumes on average 0.04–0.20 kWh/m^3. The secondary stage reaches 0.40 kWh/m^3 and the tertiary one a similar value of 0.43 kWh/m^3 (Kahrl and Roland-Holst 2008; Crawford 2009). Thus, in the case of using the three stages, a total value is reached at 1 kWh/m^3.

2.2.3 Energy Consumed for Water Used by the End User

The comfort provided in everyday life largely involves water. The water itself is used in a small part, the largest amount being hot water. Bringing it to the desired temperature and pressure again involves a significant consumption of electricity.

In the European Union, an inhabitant consumes an average of 120 L per day. Among European users, 95% use tap water for washing and body hygiene, 84% cook directly with it, 55% drink, and 10% filter it further to consume it (https://www.europarl.europa.eu/news/en/headlines/society/20181011STO15887/drinking-water-in-the-eu-better-quality-and-access).

Obviously, individual behavior, lifestyle, cultural and social factors are factors that influence the energy consumption of the residential sector (Yu et al. 2011). Seasonal alternation of climatic factors leads to higher consumption in the cold season and lower in the warm season.

Residential water heating consumes about 3.5% and 4% of the total energy demand of US and Canada, respectively (Denholm 2007; Liedl and Lubitz 2009).

Energy consumption per m3 of hot water in the USA is between 6.6 kW/h in office building, 22 kW/h in motel and 28.8 kW/h in apartment.

Even if the dishes are washed manually or automatically, the energy consumption is high or it is necessary to heat a higher amount of water or to a higher temperature. Washing clothes, manually or mechanically, involves consuming hot water preparation or pumping. The end users of a specific community can be characterized by the regional perception of cleanliness and convenience (Lin and Iyer 2007). The difference in energy consumption is quite large given the variety of materials and washing temperature. Studies indicate average values between 0.017 and 0.23 kWh/cycle/kg of clothes (Kenway et al. 2008; Yuan et al. 2010).

In his paper White (2009), he states that approximately 72% of the total energy consumption is at the end use stage, the remaining 28% representing energy consumption (kWh/m^3) with drinking water supply operations (White 2009; World Economic Outlook Database 2019).

2.3   The Economic Relationship Between Water and Energy

The two-way economic conditioning between water and energy is a sensitive issue in the implementation of the policies of sustainable management of the resources we currently have.

Reduced water and energy tariffs in general lead to modest revenues for operators so that profits do not allow investment in the modernization of production facilities or in research to modernize production technologies. At the same time, the low costs of water and energy give agriculture the opportunity to use as much water as it wants, often the extra quantities used represent losses of irrigation systems.

As we have said throughout this study, India is an example of this. Water abstractions used in agriculture amount to 85%, at the level of 2014, generally using electric pumps with low efficiency (20–35%).

Improving the efficiency of India's irrigation systems would help reduce electricity and water consumption. Suppose the efficiency of the groundwater extraction pumps is increased. In that case, a much faster depletion than this country's already expected groundwater reserves can be reached, even if the consumption of electricity would decrease or at most would keep at the same level.

In the relationship between water and energy tariffs in several countries, it is possible to see and deduce the consequences of the increase in tariffs on water consumption and energy, respectively (Fig. 2.4). Countries like Italy, Great Britain and Japan have increased electricity tariffs to lower the pressure on water reserves. At the same time, consumers are looking for solutions for the most efficient use of water, even if the gross domestic product is appreciable 4.971 trillion USD (2018) Japan, 2.828 trillion USD (2018) Great Britain or 2.072 trillion USD (2018) Italy (White 2009). For example, in Japan, gray water is used to supply toilets or irrigate green spaces. At the same time, the volumes of domestic wastewater that should be treated are reduced, thus saving energy.

Australia has one of the highest water tariffs, even if the price of energy is average, but the resources are protected by the increased tariff, so that the low resources available to this country to be used sustainably. Returning to India, it is observed that the reduced tariff for both energy and water leads to the exploitation of water resources in order to provide the necessary food for over 1.3 billion people.

A special situation is given by countries such as Saudi Arabia, Algeria or the United Arab Emirates that have low tariffs, due to their large oil reserves, and which allow them to subsidize water and energy tariffs.

Scenarios made by various organizations worldwide indicate an increase in freshwater abstraction and consumption for the next 20 years. Energy is what drives the

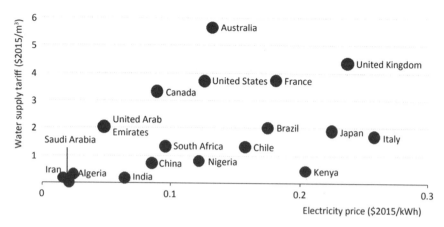

Fig. 2.4 Water and electricity tariff for selected countries, at the level of 2015, as presented by the International Energy Agency in the Water Energy Nexus 2016 report (IEA 2017)

current economic and social system. Increasing energy demand will put increasing pressure on water resources.

The demand for water and energy will clearly have an upward trend in the future. Surplus water and energy can be ensured by reducing the losses of distribution systems and increasing the efficiency of electric and thermal motors used in sampling and pumping water into the system.

Unlike energy, there is no global water market and therefore no international price for this product. The peculiarities of water supply systems, the total or partial subsidization of production costs, or the technological systems used, lead to unequal exploitation of water resources. An example of the attempt to standardize prices and eliminate disguised subsidies for certain products is the taxation of water. Agricultural products in some EU countries where water is not taxed, or energy is very cheap will no longer compete unfairly with products in countries where the aforementioned commodities are taxed or more valuable.

There are several ways in which industry and policy makers can help. First, tools such as auditing and benchmarking should be used to identify problem areas and monitor energy progress and water efficiency. Secondly, decision-makers should change the standards in relation to which drinking water and wastewater are considered important targets alongside areas such as health and the environment.

2.4 Conclusion

The pressure on water and energy resources is gaining traction. Often times the conditionality between the two is not identified, which causes major imbalances. These are either financial, high production and sales costs, or a depletion of resources, with syncope in the daily supply. The correct determination of the relationship between water and energy will lead to the establishment of a quantitative and financial value that will ensure a sustainability of the system and will not generate systemic crises that can affect the global development of society. Energy development and environmental protection policies need to be harmonized in order to meet these goals.

References

Averyt K, Fisher J, Huber-Lee A, Lewis A, Macknick J, Madden N, Rogers J, Tellinghuisen S (2011) Freshwater use by U.S. power plants: electricity's thirst for a precious resource. The Union of Concerned Scientists' Energy and Water in a Warming World initiative, November 2011 Report

Bijl DL, Bogaart PW, Kram T, de Vries BJM, van Vuuren DP (2016) Long-term water demand for electricity, industry and households. Environ Sci Policy 55:75–86. https://doi.org/10.1016/j.envsci.2015.09.005

Cheng C-L (2002) Study of the interrelationship between water use and energy conservation for a building. Energy Build 34:261–266

Civil Society Institute (2011) 200B gallons of water drawn each day for U.S. coal, nuclear power, 31 Jan

Cohen R, Nelson B, Wolff G (2004) The hidden costs of California's water supply. NRDC-Pacific Institute, Oakland, CA

Crane-Murdoch S (2010) A desperate clinch: coal production confronts water scarcity. Circle of Blue, 3 Aug

Crawford G (2009) Sustainable wastewater treatment: the intersect of water and energy, APWA. In: Conference on sustainability, March 27, USA

Dale L (2004) Electricity price and Southern California's water supply options. Resour Conserv Recycl 42:337–350

Denholm P. (2007). The technical potential of solar water heating to reduce fossil fuel use and greenhouse emissions in the United States. NREL/TP 640-41157

EEA (2019) Adaptation challenges and opportunities for the European energy system. EEA Report 1

EIA (U.S. Energy Information Administration) (2012) Analysis of the clean energy standard act of 2012. Available at: http://www.eia.gov/analysis/requests/bces12/pdf/cesbing.pdf

EPRI (2002) https://www.epri.com

Fingerman K, Berndes G, Orr S, Richter B, Vugteveen P (2011) Impact assessment at the bioenergy-water nexus. Biofuels, Bioprod Biorefin 5:375–386

Food and Agriculture Organization (2006) Introducing the international bio-energy platform. Food and Agriculture Organization, Rome, Italy

Gerbens-Leenes W, Hoekstra AY, van der Meer TH (2009) The water footprint of bioenergy. Proc Natl Acad Sci USA 106(25):10219–10223. https://doi.org/10.1073/pnas.0812619106

GWI (Global Water Intelligence) (2016) Global water market 2017. GWI, Oxford, United Kingdom

https://iea.blob.core.windows.net/assets/e4a7e1a5-b6ed-4f36-911f-b0111e49aab9/WorldEnergyOutlook2016ExcerptWaterEnergyNexus.pdf

https://unesdoc.unesco.org/ark:/48223/pf0000225741

https://voith.com/corp-en/industry-solutions/hydropower/pumped-storage-plants/guangzhou-china.html

https://www.europarl.europa.eu/news/en/headlines/society/20181011STO15887/drinking-water-in-the-eu-better-quality-and-access

https://www.ren21.net/?gclid=Cj0KCQiAmeKQBhDvARIsAHJ7mF4YNXNNPphbwYLxCaDPni1X6NBvrnOrBr73WpVK_oSNwr97H6cwoHMaAv1REALw_wcB

https://www.seia.org/initiatives/water-use-management

https://www.unicef.org/romania/stories/unicef-and-sustainable-development-goals-0

https://www.unwater.org/publications/managing-water-uncertainty-risk/

https://www.waterpowermagazine.com/news/newsihas-2015-hydropower-status-report-available-for-download-4663043

https://www.weforum.org/reports

IEA (2017) Water-Energy Nexus, IEA, Paris https://www.iea.org/reports/water-energy-nexus, License: CC BY 4.0

Jäger-Waldau A, PV Status Report (2019) EUR 29938 EN, Publications Office of the European Union, Luxembourg. ISBN 978-92-76-12608-9. https://doi.org/10.2760/326629, JRC118058

Kahrl F, Roland-Holst D (2008) China's water–energy nexus. Water Policy 10:51–65

Kenway SJ, Priestley A, Cook S, Seo S, Inman M, Gregory A, Hall M (2008) Energy use in the provision and consumption of urban water in Australia and New Zealand. CSIRO: Water for a Healthy Country National Research Flagship

Kneppers B, Birchfield D, Lawton M (2009) Energy-water relationships in reticulated water infrastructure systems. WA7090/2, pp 1–31

Lance F (2004) Low water consumption: a new goal for coal. Environmental health perspectives, April

Li G, Ma SQ, Xue XX, Yang SC, Liu F, Zhang YL (2021) Life cycle water footprint analysis for second-generation biobutanol. Bioresour Technol 333, Article Number 125203. https://doi.org/10.1016/j.biortech.2021.125203

Liedl CM, Lubitz WD (2009) Comparing domestic water heating technologies. Technol Soc 31:244–256

Lin J, Iyer M (2007) Cold or hot wash: technological choices, cultural change and their impact on clothes—washing energy use in China. Energy Policy 35:3046–3052

Luck H, Tsai S, Chung J, Clemente- X, Ghazarian M, Revelo XS, Lei H, Luk CT, Shi SY, Surendra A, Copeland JK, Jennifer A, Prescott D, Rasmussen BA, Melissa Hui Chang Y, Engleman EG, Girardin SE, Lam TKT, Croitoru K, Dunn S, Philpott DJ, Guttman DS, Woo M, Winer S, Winer DA (2015) Regulation of obesity-related insulin resistance with gut anti-inflammatory agents. Cell Metab 21:527–542

Maina FZ, Rhoades A, Siirila-Woodburn ER, Dennedy-Frank PJ (2022) Projecting end-of-century climate extremes and their impacts on the hydrology of a representative California watershed. Hydrol Earth Syst Sci 26(13):3589–3609. https://doi.org/10.5194/hess-26-3589-2022

Martinelli LA, Filoso S, Aranha CDB, Ferraz SFB, Andrade TMB, Ravagnani EDC, Coletta LD (2013) Water use in sugar and ethanol industry in the state of São Paulo (Southeast Brazil). J Sustain Bioenergy Syst 3:135–142

NDRC (2014). Power plant cooling and associated impacts: the need to modernize U.S. power plants and protect our water resources and aquatic ecosystems

NETL (2010) Water vulnerabilities for existing coal-fired power plants, DOE/NETL-2010/1429

Pitorac L, Vereide K, Lia L (2020) Technical review of existing Norwegian pumped storage plants. Energies 13(18):4918. https://doi.org/10.3390/en13184918

Plappally AK, Lienhard JHV (2012) Energy requirements for water production, treatment, end use, reclamation, and disposal. Renew Sustain Energy Rev 16(7):4818–4848

Rogner M, Troja N (2018) The world's water battery: pumped hydropower storage and clean energy transition. IHA, London, UK

Spang ES, Moomaw WR, Gallagher KS, Kirshen PH, Marks DH (2014) The water consumption of energy production: an international comparison. Environ Res Lett 9(10)

Stokes J, Horvath A (2006) Life cycle energy assessment of alternative water supply systems. Int J Life Cycle Anal 2006(11):335–343

The United Nations World Water Development Report (2014) Water and energy. UNESCO, Paris

Wada Y, Flörke M, Hanasaki N, Eisner S, Fischer G, Tramberend S, Satoh Y, van Vliet MTH, Yillia P, Ringler C, Burek P, Wiberg D (2016) Modeling global water use for the 21st century: the Water Futures and Solutions (WFaS) initiative and its approaches. Geosci Model Dev 9:175–222. https://doi.org/10.5194/gmd-9-175-2016

Wendy W, Travis L, Bevan G-S (2012) Burning our rivers: the water footprint of electricity, river network report, April 2012
White L (2009) Energy–water nexus: critical actions in a climate change world. Western States Water Council; 1
World Economic Outlook Database (2019) Gross domestic product, current prices, U.S. dollars. International Monetary Fund, April
WWAP (2009) World water assessment programme. The United Nations world water development report 3. Water in a changing world. http://www.unesco.org/new/fileadmin/MULTIMEDIA/HQ/SC/pdf/WWDR3_Facts
WWAP (2012) The United Nations world water development report 4: managing water under uncertainty and risk. UNESCO, Paris
www.waterpowermagazine.com/features/featurea-capable-and-flexible-technology-8361219/
Yu B, Zhang J, Fujiwara A (2011) Representing in-home and out-of-home energy consumption behavior in Beijing. Energy Policy
Yuan C, Liu S, Wu J (2010) The relationship among energy prices and energy consumption in China. Energy Policy 38:197–207
Zarfl C, Lumsdon AE, Berlekamp J, Tydecks L, Tockner K (2015) A global boom in hydropower dam construction. Aq Sci 77:161–170

Chapter 3
Energy Sectors that Use Water

Abstract Electricity is produced in different ways. Cost price varies depending on the technological process. Water is the resource that most use. This chapter shows the consumption of water necessary for electricity depending on technology. The climate evolution, as well as conditioning imposed regarding environmental protection will generate new restrictions related to water usage in electricity production.

Keywords Electricity technology · Water

All economic sectors need water to operate. Agriculture, industry and most energy production cannot function if they do not have access to the water source. It is the climate that determines the availability of water and the seasonality of water demand, and the latter depends on population density and is connected to economic activities (United Nations 2012; Cojanu and Helerea 2021; Grubert and Marshall 2022; de Oliveira et al. 2022).

Electricity production is a branch that uses, among others, fuel and important water resources. The impact of this industry on water is both quantitative and qualitative, depending on how electricity is produced. Hydropower plants do not have a major quantitative impact on water resources, producing only a redistribution over time. Nuclear power plants that use large volumes of water to cool reactors produce a thermal deterioration of water with a direct impact on aquatic biodiversity. Coal, oil, gas or other thermal power plants use water for washing and briquetting coal, for steam production, etc. It is true that water consumption is not very high (Fig. 3.1).

Seasonal or annual water samples show a series of variations and trends, depending on the climatic variations, the number of users in a certain hydrographic area but especially by the industrial evolution and the technological performance achieved in the collection, transport, use and reuse of the water resource (EEA 2012a, b).

All economic sectors need water to run their businesses. Agriculture, industry and most forms of energy production cannot function if water is not available. The analysis index called Water Exploitation Index Plus is based on two important factors. Climate control gives water availability and seasonality, and water demand, in turn, depends on population density and related economic activities. In the Mediterranean biogeographical region, these two factors coincide, leading to high values of indicators. In other biogeographical regions, water stress often occurs in areas associated

Fig. 3.1 Annual water use by sector, 2002–2012. *Source of data* https://www.eea.europa.eu/data-and-maps/indicators/use-of-fre shwater-resources-2/assessment-1

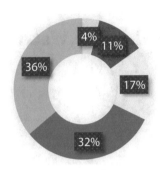

with high population density, except for those with dry summer conditions over time. At certain times of the year and in areas with high indicator values, certain economic sectors become the main drivers of water demand. For example, agriculture leads to high values of indicators in spring and summer, and autumn and winter are the peak seasons for electricity use. The industry consumes less water during the summer compared to other seasons (Ivits et al. 2012).

3.1   Water Sampling for Energy Production

Water sampling for energy production through hydropower units is considered non-water consuming, which means that all water taken is returned to the environment. Water is also used by thermal power plants for steam production and thus cooling. Most of the water used during the cooling process is returned to the environment, some quantities being lost by evaporation. However, the water resource taken and used also has a quantitative and an ecosystem side, of great importance. The realization of large accumulations of water, derivatives and aqueducts lead to changes, and might be irreversible, in terms of fauna, flora and landscape. At the same time, the storage of water over long periods of time, months or even years, produces a change in its physico-chemical composition. Due to stagnation, lack of light, etc., water changes its original composition, causing alterations by increasing the amounts of CO_2, H_2S or CH_3.

Under these conditions, we can say that hydropower is not a quantitative consumer of water resources but involves a qualitative consumption with important ecosystem implications, which should be found in energy production costs. Not only fresh water is used in the process of obtaining electricity. For example, water from coastal and estuarine areas is also used for cooling purposes, accounting for about 25% of the total amount of water used to produce electricity. Electricity is also obtained from the use of hydraulic energy from waves or tides in the coastal area, with low quantitative implications but which influence landscape or wildlife in certain situations.

The mainland uses almost 65% of its total water supply for electricity, gas, steam and air conditioning, followed by the Atlantic region (15%) and the Mediterranean region (13%).

In general, after the 1990s, water abstractions at the European level are declining. For example, the industrial sector has improved its water use efficiency, which has led to a significant decrease in the quantities used by 27%. Agriculture also has a 22% decrease in water abstraction, but agriculture is still the sector with the highest demand for water. A significant increase in water use in agriculture (140%) was observed in Turkey between 1990 and 2013. While in Romania, the destruction and non-use of large irrigation systems or the charging of water for irrigation has led to a significant decrease in water abstractions in this regard.

Electricity water use has fallen by 11% since the 1990s, indicating a mostly steady trend since the year 2000. Investments in the refurbishment of the current hydropower plants, the decrease in consumption due to going through periods of economic crisis with major repercussions on the industry, may be some of the causes of this decrease. The evaluation of the quantities of water taken from the hydric environment can be done in different ways. One of them is the use of indicators that add up to several factors.

The WEI+ provides a measure of the total water use as a percentage of the renewable freshwater resources for a given territory and time scale.

The WEI+ is an advanced and geo-referenced implementation of the WEI. It quantifies how monthly or seasonally abstracted water and how much water is returned after use to the environment via basins. The difference between water abstraction and return is known as water use. WEI + values are given as percentages, i.e., water use percentage of renewable water resources. Absolute water volumes are presented as millions of cubic meters (million m^3 or hm^3).

Monitoring water use efficiency is important for the protection, conservation and enhancement of the EU's natural capital. It also contributes to improving resource efficiency, which is included as an objective of the EU's 7th EAP to 2020.

The WEI+ is a water scarcity indicator that provides information on the level of pressure that human activity exerts on the natural water resources of a territory. This helps to identify those areas prone to problems related to water stress (Faergemann 2012). Implementing the WEI+ at spatial (e.g., sub-basin or river basin) and temporal (monthly or seasonal) scales, which are finer than annual averages at the country scale, is to better capture the balance between renewable sources water resources and water use.

The WEI+ is an advanced version of the water exploitation index. It is geo-referenced and developed for use on a seasonal scale. It also considers water abstraction (gross) and return (net abstraction) to reflect water use.

In 2011, a technical working group, developed under the Water Framework Directive Common Implementation Strategy, proposed the implementation of a regionalized WEI+. This differed from the previous approach by enabling the WEI+ to depict more seasonal and regional aspects of water stress conditions across Europe. The Water Directors approved this proposal in 2012 as one of the awareness-raising indicators.

The regionalized WEI+ is calculated according to the following formula (Eq. 3.1):

$$\text{WEI} + = (\text{abstractions} - \text{returns})/\text{renewable freshwater resources} \qquad (3.1)$$

Renewable freshwater resources are calculated as "ExIn + P − Eta − ΔS" for natural and semi-natural areas, and as "outflow + (abstraction − return) − ΔS" for densely populated areas.

where: ExIn = external inflow; P = precipitation; ETa = actual evapotranspiration; ΔS = change in storage (lakes and reservoirs); outflow = outflow to downstream/sea.

It is assumed that there are no pristine or semi-natural river basin districts or sub-basins in Europe. Therefore, the formula 'outflow + (abstraction − return) − ΔS' is used to estimate renewable water resources.

Climatic data were obtained from the EEA Climatic Database, which was developed based on the ENSEMBLES Observation (E-OBS) Dataset (Haylock et al. 2008). The State of the Environment database was used to validate the aggregation of the E-OBS data to the catchment scale.

Streamflow data have been extracted from the EEA Waterbase—Water Quantity database. This database does not have sufficient spatial and temporal coverage yet. To fill the gaps, Joint Research Centre (JRC) LISFLOOD data have been integrated into the streamflow data (Burek et al. 2013). The streamflow data cover Europe, in a homogeneous way, for the years 1990–2015 on a monthly scale.

Once the data series are complete, the flow linearization calculation is implemented, followed by a water asset accounts calculation, which is done to fill the data for the parameters requested for the estimation of renewable water resources. The computations are implemented at different scales independently, from sub-basin scale to river basin district scale.

Overall, annually reported data are available for water abstraction by source (surface water and groundwater) and water abstraction by sector with temporal and spatial gaps. Gap-filling methods are applied to obtain harmonized time series.

No data are available at the European scale on "Return". Urban wastewater treatment plant data, the European Pollutant Release and Transfer Register (E-PRTR) database, Eurostat population data, JRC data on the crop coefficient of water consumption, satellite observed phenology data have been used as proxy to quantify the water demand and water use by different economic sectors. Eurostat tourism data and data on industry in production have been used to estimate the actual water abstraction and return on a monthly scale. Where available, state of the environment and Eurostat data on water availability and water use have also been used at aggregated scales for further validation purposes (Eurostat 2013; Essex et al. 2004; Gössling et al. 2012).

Once water asset accounts are implemented according to the United Nations System of Environmental Accounting Framework for Water (2012), the necessary parameters for calculating water use and renewable freshwater water resources are harvested.

Following this, bar and pie charts are produced with static and dynamic maps.

Data are very sparse on some parameters of the WEI+. For instance, current streamflow data reported by the EEA member countries to the WISE SoE—Water Quantity database do not have sufficient temporal or spatial coverage to provide a strong enough basis for estimating renewable water resources for all of Europe. Such data are not available elsewhere at the European level either. Therefore, JRC LISFLOOD data are used intensively as surrogates.

Data on water abstraction by economic sectors have better spatial and temporal coverage. However, the representativeness of data for some sectors is also poor, such as the data on water abstraction for mining. In addition to the WISE SoE—Water Quantity database, intensive efforts to compile data from open data sources such as Eurostat, OECD, Aquastat (FAO) and national statistical offices have also been made.

Quantifying water exchanges between the environment and the economy is, conceptually, very complex. A complete quantification of the water flows from the environment to the economy and, at a later stage, back to the environment, requires detailed data collection and processing, which have not been done at the European level. Thus, reported data must be used in combination with modeling to obtain data that can be used to quantify such water exchanges to develop a good approximation of "ground truth". However, the most challenging issue is related to water abstraction and water use data, as the water flow within the economy is quite difficult to monitor and assess given the current lack of data availability. Therefore, several interpolation, aggregation or disaggregation procedures must be implemented at finer scales, with both reported and modeled data. Main consequences of data set uncertainty are the following:

1. The Danube River basin is accounted for as a single district in Ecrins, so it aggregates a lot of regional and national information.
2. The water accounts and WEI+ results have been implemented in the EEA member and Western Balkan countries (https://www.eea.europa.eu/data-and-maps/indicators/direct-losses-from-weather-disasters-3/assessment-2) (EEA 2009). However, regional data availability was an issue for some river basins (e.g., in Cyprus, the Jarft in Poland, Northwest and North Eastern River basins in the United Kingdom, the Kymijoki river basin in the Gulf of Finland, Gran Canarias of Spain and some Icelandic and Turkish river basins). These had to be removed from the assessment.

The EU's 7th EAP to 2020 aims to ensure the protection, conservation and enhancement of the EU's natural capital and improve resource efficiency. Monitoring the efficiency of water use in different economic sectors at national, regional and local levels is necessary to achieve this. The WEI is part of the set of water indicators published by several international organizations, such as the United Nations Environment Program (UNEP), the Organization for Economic Co-operation and Development (OECD), Eurostat and the Mediterranean Blue Plan. There is an international consensus about using this indicator to assess the pressure of the economy on water resources, i.e., water scarcity.

The WEI+ is an advanced version of the WEI, which better addresses regional and seasonal aspects of water scarcity. In addition, it also takes water use (water abstraction minus water returned) into account. The indicator describes how total water use exerts pressure on water resources. It identifies areas (e.g., sub-basins or river basins) with high abstraction levels on a seasonal scale in relation to the available resources and prone to water stress. Changes in WEI+ values allow analyses of how changes in water use affect freshwater resources, i.e., by putting them under pressure or by making them more sustainable.

Related policy documents

Multiple documents have been elaborated regarding the sustainable use of the planet's resources. At the level of the EU, several indicators were set that member countries must comply with. One of these documents is decision No. 1386/2013/Eu of the European Parliament and of the Council of 20th of November 2013, on a General Union Environment Action Program to 2020 "Living well, within the limits of our planet". This program is intended to help guide EU action on the environment and climate change up to and beyond 2020 based on the following vision: "In 2050, we live well, within the planet's ecological limits. Our prosperity and healthy environment stem from an innovative, circular economy where nothing is wasted and where natural resources are managed sustainably, and biodiversity is protected, valued, and restored to enhance our society's resilience. Our low-carbon growth has long been decoupled from resource use, setting the pace for a safe and sustainable global society".

Hydrological aspects regarding the management of the water resources were addressed, such as the challenge of water scarcity and droughts in the European Union.

Several related documents regarding this include:

- EC (2007). Communication from the Commission to the Council and the European Parliament, Addressing the challenge of water scarcity and droughts in the European Union. Brussels, 18.07.07, COM (2007) 414 final.
- Roadmap to a Resource Efficient Europe COM (2011) 571
- Communication from the Commission to the European Parliament, the Council, the European Economic and Social Committee, and the Regions Committee. Roadmap to a Resource Efficient Europe. COM (2011) 571
- Water Framework Directive (WFD) 2000/60/EC
- Water Framework Directive (WFD) 2000/60/EC: Directive 2000/60/EC of the European Parliament and of the Council of 23 October 2000 establishes a framework for Community action in water policy.

Through the direct use of a significant amount of water, the hydropower sector is at risk of not meeting its water demand due to climate variability and thus diminished water resources throughout the year or over a longer (multiyear) period. At the same time, some extreme natural phenomena can lead to economic losses in the energy production system.

In the EEA member countries (EEA-33), the total reported economic losses caused by weather and climate-related extremes throughout 1980–2017 amounted to approximately 453 billion Euro (in 2017 Euro values).

Average annual economic losses in the EEA member countries varied between 7.4 billion Euro over the period 1980–1989, 13.4 billion Euro (1990–1999) and 14.0 billion Euro (2000–2009). Between 2010 and 2017, average annual losses were around 13.0 billion Euro. This high variability makes the analysis of historical trends difficult, since the choice of the studied years heavily influences the trend outcome and thus the conclusions.

Over time, the observed variations in reported economic losses are difficult to interpret since a large share of the total deflated losses has been caused by a small number of events. Specifically, more than 70% of economic losses were caused by less than 3% of all unique registered events.

In the EU Member States (EU-28), disasters caused by weather and climate-related extremes accounted for 83% of the monetary losses over the period 1980–2017. Weather and climate-related losses amounted to 426 billion Euro (at 2017 values).

The most expensive climate extremes in the EU Member States include the 2002 flood in Central Europe (over 21 billion Euro), the 2003 drought and heat wave (almost 15 billion Euro), and the 1999 winter storm Lothar and October 2000 flood in Italy and France (both 13 billion Euro), all at 2017 values (https://www.eea.europa.eu/data-and-maps/indicators/direct-losses-from-weather-disasters-3/assessment-2).

The losses are induced by long periods of meteorological and therefore hydrological drought, which leads to small volumes of water that can be processed by hydropower plants and transformed into energy. At the same time, periods of excess water, flood waves, are water events that do not produce an excess of energy, but on the contrary, certain losses. Now, these losses can be considered as such in the short term, and I mean the volumes of water to be transited without being processed (ecological water flow). In the case of applying the rules of safe operation of dams (high water flow transit), the increased turbidity of water makes it unusable in the case of thermal power plants. In the long run, the flood waves passing through the accumulation lakes can clear the accumulation lakes formed behind the dams, increasing their useful volume. However, the biggest losses are generated by the dams' structural damage or certain functional components.

According to data on natural disasters in the European Environment Agency (EEA) member countries, between 1980 and 2017—from NatCatSERVICE in Munich Re—climate and meteorological extremes accounted for about 81% of total losses caused by extreme natural events.

Specifically, weather and climate losses amounted to €453 billion (In the 2017 currency value), averaging €12 billion per year and €79,200 per square kilometer or €811 per capita. The cumulative losses broken down over the analyzed period are equal to almost 3% of the GDP of all EEA member countries in 2017. Overall, about 35% of total losses were insured, although the percentage of insured losses ranged from 1% in Romania and Lithuania to 70% in the United Kingdom.

Unfortunately, 90,325 victims were registered during this period (www.munichre.
com/natcatservice).

The reported economic losses mainly reflect direct short-term damage to certain
assets. The loss of human life, cultural heritage or ecosystem services is not part of
the estimate. Their non-quantified values can significantly increase the previously
presented value of the losses caused. The weather and climate losses distribution
among the 33 EEA member states are uneven. The largest total economic losses in
absolute terms (in order of rank) were recorded in Germany, Italy and France. The
highest per capita losses were recorded in Switzerland, Denmark and Austria, while
those per square kilometer were recorded in Switzerland, Luxembourg and Germany.
Most of the total losses in terms of cumulative GDP were recorded in Croatia,
the Czech Republic and Hungary. The three countries least affected in absolute
terms were Liechtenstein, Malta and Iceland. In relative terms (per capita), the least
affected countries were Turkey, Estonia and Malta (https://www.eea.europa.eu/data-
and-maps/indicators/direct-losses-from-weather-disasters-3/assessment-2). In terms
of the loss of cumulative GDP as a share, the least affected countries were Liechtenstein, Iceland and Estonia. The 39 largest events accounted for about half of the
losses.

It is important to understand the extent to which the observed increase in global
losses over the last decades is due to changing climatic conditions rather than
other factors. According to the IPCC's AR5, the increasing exposure of people
and economic assets to weather and climate disasters has been the main cause of
their long-term growth in economic losses. Available studies on the economic losses
caused by floods and storms in Europe suggest that the increase in losses is mainly
due to population growth, economic wealth and developments in hazardous areas, but
the significant increase in heavy rainfall in some parts of Europe also played a role.
There is evidence that improving flood protection and prevention has helped reduce
losses over time in some cases. Globally, "attribution science" has made significant
progress in recent years in assessing whether global climate change has affected the
chances of a major weather event. A recent analysis of these studies shows that,
globally, the vast majority of the heat waves analyzed and most droughts and rain
and flood events were considered more likely and/or more severe as a result of global
climate change. Attributing storms and (other) small-scale events to global climate
change is much more difficult, mainly due to their poor representation in climate
models.

For the period 1980–2017, the economic losses generated by all natural disasters in the EEA member countries were 557 billion Euro, and the insured losses
were approximately 162 billion Euro (in 2017 values). About 63% of all economic
losses were the result of meteorological and hydrological events, while most deaths
were caused by heat waves. The sudden rise of the deaths is strongly affected
by the heat waves of 2003, where about 68,000 deaths were reported as excess
mortality. In Europe, the economic losses caused by climate and climate extremes
have varied substantially over time. The average annual economic loss (correlated
with inflation) was around 7.4 billion Euro per year in the 1980s, 13.4 billion
Euro in the 1990s and 14.0 billion Euro per year in the 2000s (2000–2009). In

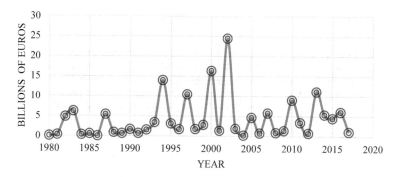

Fig. 3.2 Economic damage caused by hydrological event EU-28, 1980–2017. *Data source* https://www.eea.europa.eu/data-and-maps/indicators/direct-losses-from-weather-disasters-3/ass essment-2

the period 2010–2017, the average annual economic loss amounted to approximately 13.0 billion Euro (https://www.eea.europa.eu/data-and-maps/indicators/dir ect-losses-from-weather-disasters-3/assessment-2). However, the pattern found in the recorded loss is covered by high variability: about 3% of events—some of which affect more than one country—account for about 75% of total deflated losses. In contrast, about three-quarters of the recorded events accounted for only 0.7% of total losses. Increasing economic wealth has a major effect on annual losses.

The losses generated by the extreme hydrological events at the level of the European Union register annual average values that do not generally exceed 2.5–3 billion Euro, but there are also years in which the losses reach the threshold of 10, 15, 25 billion Euro. In the analyzed period, 1980–2017, the high value of 2002—24.45 billion Euro economic losses is highlighted, followed closely by 2000—16.35 billion Euro and 1994—13.98 billion Euro (Fig. 3.2) (https://www.eea.europa.eu/data-and-maps/indicators/direct-losses-from-weather-disasters-3/assessment-2, www.munichre.com/natcatservice) (OECD 2014; JRC 2015; IRDR 2015; OEIWG 2016).

3.2 Conclusion

Water is the resource that is directly or indirectly involved in electricity production. The current and future challenges will be in the optimal allocation of this resource in the context of increasing electricity needs but especially in the current climatic conditions. These have led to the emergence of extreme events (droughts or floods), which the energy industry in general has never experienced before. This has resulted in very high economic and human losses. Global or continental organizations have adopted various strategies to mitigate the negative effects of global climate change, and to adapt to current energy requirements. The challenges ahead will be all the greater as greater emphasis is placed on the regeneration of natural ecosystems and environmental protection in general.

References

Burek P, Kniff van der J, Roo de A (2013) LISFLOOD, distributed water balance and flood simulation model revised user manual 2013. In: European Commission joint research centre institute for the protection and the security of the citizen, Luxembourg. ISBN: 9789279331909 9279331906

Cojanu V, Helerea E (2021) On the water-energy relationship in the case of water supply systems. In: International conference on applied and theoretical electricity (ICATE) book series. https://doi.org/10.1109/ICATE49685.2021.9464941

de Oliveira GC, Bertone E, Stewart RA (2022) Challenges, opportunities, and strategies for undertaking integrated precinct-scale energy-water system planning. Renew Sustain Energy Rev 161, Article Number 112297. https://doi.org/10.1016/j.rser.2022.112297

EEA. https://www.eea.europa.eu/data-and-maps/indicators/use-of-freshwater-resources-2/assessment-1

EEA. https://www.eea.europa.eu/data-and-maps/indicators/direct-losses-from-weather-disasters-3/assessment-2

EEA (2009) Water resources across Europe—confronting water scarcity and drought, EEA Report No 2/2009. European Environment Agency, Copenhagen

EEA (2012a) European waters—current status and future challenges—synthesis. EEA Report No 9/2012, European Environment Agency.

EEA (2012b) Water resources in Europe in the context of vulnerability. EEA Report No 11/2012, European Environment Agency

Essex S, Kent M, Newnham R (2004) Tourism development in Mallorca: is water supply a constraint? J Sustain Tour 12(1)

Eurostat (2013) Tourism industries—economic analysis. Eurostat statistics explained, statistics in focus 32/2013

Faergemann H (2012) Update on water scarcity and droughts indicator development, May 2012, presented at the Water Director's Meeting, 4–5 June 2012, Denmark

Gössling S, Peeters P, Michael Hall CM, Ceron JP, Dubois G, Lehmann LV, Scott D (2012) Tourism and water use: supply, demand, and security. Int Rev Tour Manage 33(1):1–15

Grubert E, Marshall A (2022) Water for energy: characterizing co-evolving energy and water systems under twin climate and energy system nonstationarities. Wiley Interdiscip Rev Water 2(9), Article Number 1576. https://doi.org/10.1002/wat2.1576

Haylock MR, Hofstra N, Klein Tank AMG, Klok EJ, Jones PD, New M (2008) A European daily high-resolution gridded data set of surface temperature and precipitation for 1950–2006. J Geophys Res 113(D20)

IRDR (2015) Guidelines on measuring losses from disasters: human and economic impact indicators. DATA project report no. 2, Integrated research on disaster risk programme, Beijing

Ivits E, Cherlet M, Mehl W, Sommer S (2012) Ecosystem functional units characterized by satellite observed phenology and productivity gradients: a case study for Europe land resource management unit, EC Joint Research Centre (JRC), Via E. Fermi 1, 21020 Ispra (VA) Italy. Received 12 April 2012, Revised 8 Nov 2012, Accepted 13 Nov 2012, Available online 20 Dec 2012. https://doi.org/10.1016/j.ecolind.2012.11.010

JRC (2015) Guidance for recording and sharing disaster damage and loss data: towards the development of operational indicators to translate the Sendai framework into action. JRC science and policy reports, joint research centre, institute for the protection and security of the citizen and the EU expert working group on disaster damage and loss data, Ispra

OECD (2014) Improving the evidence base on the costs of disasters: towards an OECD framework for accounting risk management expenditures and losses of disasters. In: 4th Meeting of the OECD high level risk forum, GOV/PGC/HLRF (2014)8, OECD

OEIWG (2016) Report of the open-ended intergovernmental expert working group on indicators and terminology relating to disaster risk reduction (Geneva, 29–30 Sept 2015, 10–11 Feb 2016 and 15 & 18 Nov 2016)

United Nations (2012) System of environmental-economic accounting for water. United Nations, New York

www.munichre.com/natcatservice

Chapter 4
The Context for the Development of the Global Hydropower Sector

Abstract The hydropower potential of the rivers was among the first to be exploited in terms of energy. Thus, the scale of hydropower development has increased with industrial and technological development. Depending on the physical-geographical and flow characteristics, dams and hydropower plants have been set up, which provide significant amounts of energy. Obviously, this potential is exploited differently from country to country, but the trend is one of growth in general. The value presentation of the exploited potential by countries helps to understand the need and contribution of this energy supply branch.

Keywords Hydropower · Production capacities

The hydropower sector is the basic alternative to electricity production compared to the one that traditionally uses fossil fuels. Hydropower generates small amounts of greenhouse gases and other air pollutants (Klimpt et al. 2002). Even though the pressures on ecosystems exerted by large hydropower constructions have been taken into discussion lately, the company has high expectations in this field. Moreover, in the context of the restructuring of the electricity sector, government markets and policies can favor less polluting sectors.

4.1 Values of the Hydropower Capacities Installed in the World

China was the world's largest hydropower capacity of around 356 GW in 2019. This is followed by Brazil 109.06 GW, US 102.75 GW, Canada 81.39 GW, Japan, India and Russia 50 GW, Norway 32.67 GW, Turkey 28.50 GW, France 25.56 GW, Italy 23 GW, Spain 20 GW, Switzerland and Vietnam 17 GW, Sweden 16 GW, Venezuela and Austria with 15 GW, Mexico, Iran and Colombia with 12 GW. The rest of the world also accounts for 274.43 GW or 76% of China's installed capacity (Manzano-Agugliaro et al. 2021).

The capacity to store water through re-pumping is also increasingly significant, ensuring the flexibility of energy systems, thus ensuring energy at peak times. In

© The Author(s), under exclusive license to Springer Nature Switzerland AG 2023
D. C. Diaconu, *Force Majeure in the Hydropower Industry*,
https://doi.org/10.1007/978-3-031-27402-2_4

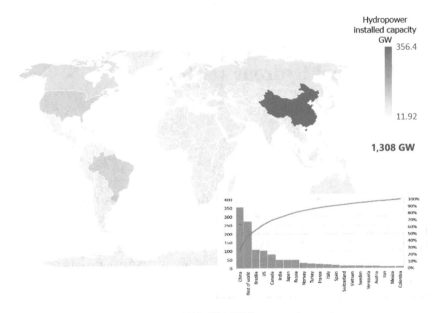

Fig. 4.1 Global hydropower capacities, 2019 (IHA 2020)

this ranking, China is the highest capacity country 30 GW, followed by Japan 27.6 GW, US 22.9 GW, Italy 7.6 GW, France 7.0 GW, Germany 6.8 GW, Spain 6.2 GW, Austria 5.5 GW, India 4.8 GW and South Korea with 4.7 GW. The rest of the countries cumulate 37.2 GW, so the total value reaches 160.3 GW (IHA 2020) (Fig. 4.1).

Why this evolution of the hydropower sector in the world?

At the global level, there is an upward trend in the value of production capacity. This is mainly due to the high demand for electricity, the profit generated and the technological evolution in the construction of dams and hydropower generators. Last but not least, the concern for the reduction of greenhouse gases and especially carbon concentrations has led to increased investment in this energy field.

In 2015, the installed power worldwide reached the value of 1211 GW. It currently exceeds 1308 GW, 8% more, according to data published by the International Hydropower Association, 2020.

First, the current production price is among the lowest, with the average cost being 0,05 USD/kWh. For example, the sale prices of electricity (including taxes) for household consumers in the first half of 2021, according to Eurostat and at the level of Europe is 0.265 USD/kWh (https://ec.europa.eu/eurostat/statistics-explai ned/index.php?title=Electricity_price_statistics). Availability of hydropower at peak times when conventional energy (coal, gas) or renewable energy such as wind or solar energy cannot provide it. Water-energy developments in the river basins play a complex role, protecting against floods and droughts, thus ensuring that volumes of water are regulated over time and space. It provides many direct and horizontal industrial jobs, exceedingly over 1,8 million jobs worldwide. It is a low carbon, which is generated during the construction phase and less during operation. It improves shipping routes, the transport network, and economic and political relations between countries. Last but not least, it ensures the development of the recreational and fisheries sector.

4.2 Hydropower, Renewable Energy?!

Hydropower is the most developed energy sector among renewable sources. The share of renewables is currently around 25.6% of all energy produced, with the remaining 74.4% still made from fossil and nuclear fuels. Renewable energy is energy produced by harnessing the energy potential of water (15.9% of global energy production), wind (4.6%), biomass and waste (2.5%), sun (2.1%) and other sources (0.5%).

However, these renewable sources of electricity also generate carbon emissions into the atmosphere and other greenhouse gases but much less than fossil fuels (Chen 2021).

Compared to coal-fired power generation in thermal power plants, they emit 820 gr CO_2-eq./kWh into the atmosphere, biomass energy production sources emit 230 gr CO_2-eq./kWh into the atmosphere, solar installations 48 gr CO_2-eq./kWh, geothermal 38 gr CO_2-eq./kWh, hydropower 18.5 gr CO_2-eq./kWh and wind 11 gr CO_2/kWh (IEA 2019).

In this context, the same amount of electricity produced from coal is lower by around 10% when it reports greenhouse gas emissions from the hydropower sector.

However, the share of this energy sector is limited to a specific value, given the hydropower potential of rivers and the possibility of new accumulation and retention tanks.

The development of this area has been accelerating since 1950 when the energy requirement of the world industry in reconstruction was very high since the Second World War. A further increase in growth has been observed since 2000, when demands for clean energy have been made, with an attempt to reduce dependence on fossil fuels.

Asia is the continent with the most considerable increase in newly installed production capacities, so China in 2018 has more than 8540 MW of installed capacity, leading in the world. This is possible thanks to the vast water resources at its disposal and the economic and political regime, making it possible to invest heavily in this sector (Gutierrez et al. 2019).

South America is the second region, where Brazil, with 3866 MW installed power, is the second-largest in the world. The two regions are followed by Central and South Asia, where Pakistan with 2487 MW and Turkey with 1085 MW stand out, Africa where Angola with 668 MW, and Central and North America where Canada has 401 MW.

Romania has not given new large hydropower capacities in recent years, with investments in this sector being made preweighting by private investors in microhydropower plants. However, there are European countries that have completed new investments, such as Slovenia (45 MW), Switzerland (26 MW), Spain (17 MW), Serbia (15 MW), Bosnia and Herzegovina (9 MW), the Czech Republic, and Portugal (4 MW) and France (2 MW) (IHA 2019).

Romania is ranked among European countries with the 12-6328 MW hydropower plant installed capacity. The top three seats are Norway (32,256 MW), Turkey (28,358) and France (25,519 MW) at the 2018 level (Table 4.1) (https://ec.europa.eu/eurostat/statistics-explained/index.php?title=Electricity_price_statistics).

The hydropower sector has not experienced significant growth at the European level over the last ten years, with investments focusing more on renewable energies such as solar and wind (Strielkowski et al. 2021). However, the decommissioning of

Table 4.1 Hydropower capacities installed by country at 2018

No.	Country	Total available capacity (MW)	No.	Country	Total available capacity (MW)
1	Norway	32,256	22	Croația	2141
2	Turkey	28,358	23	Iceland	2086
3	France	25,519	24	Albania	2020
4	Italy	22,926	25	Lithuania	1576
5	Spain	20,378	26	Slovenia	1524
6	Switzerland	16,948	27	Belgium	1427
7	Sweden	16,466	28	Luxembourg	1330
8	Austria	14,535	29	Latvia	1016
9	Germany	11,258	30	North Macedonia	674
10	Portugal	7347	31	Montenegro	658
11	Ukraine	6785	32	Ireland	529
12	Romania	6328	33	Belarus	97
13	United Kingdom	4712	34	Greenland	91
14	Greece	3396	35	Moldova	76
15	Finland	3236	36	Kosovo	68
16	Bulgaria	3129	37	Hungary	56
17	Serbia	2932	38	Andorra	45
18	Slovakia	2522	39	Faroe Islands	39
19	Bosnia and Herzegovina	2513	40	Netherlands	37
20	Poland	2353	41	Liechtenstein	35
21	Czech Republic	2268	42	Denmark	9

Data source IHA (2020).

nuclear power plants has led to hydropower production to ensure the energy system's flexibility to remain safe and without blackouts.

They are energy systems where this is more difficult to do, and a quick solution is the interconnection of the energy systems of European countries. For example, Germany's energy system is interconnected with 12 other energy supply systems. Turkey is investing in new hydropower projects to reduce its dependence on energy imports as much as possible.

Austria has recently completed two pumping and storage systems to ensure flexibility in its energy system, given the wind and solar energy entry into the system.

France also wants to do the same, with storage-pumping systems in the design and execution of around 10 GW by the end of 2035.

Slovenia makes the fourth settlement on the Sava River, cascading after Boštanj, Blanca and Krško (https://www.hydropower.org/news/iha-releases-2019-hydropower-status-report-charting-growth-in-renewable-hydro).

4.3 Conclusion

Hydropower is considered to be a green source of energy, a renewable source given the water cycle in nature. Renewable energy sources include hydropower, bioenergy, thermal, geothermal, wind, photochemical, photoelectric, tidal, wave, and solar energy. It excludes energy from fossil fuel sources (oil, coal and natural gas) (TREIA 2015). Its exploitation has generated hydrotechnical arrangements of different sizes, many of them involving even several neighboring river basins. Energy production costs being some of the lowest, if not the lowest compared to other types (Güney 2019, 2021).

Over time, however, this type of electricity generation has produced major changes, and most often irreversible, in aquatic ecosystems. Quantifying environmental losses at their fair value can lead to an increase in the cost of production and a change in the perspective on the exploitation of this resource (Gheorghe et al. 2017).

At the same time, the depletion of sites where dams can be built, as well as the physical and moral obsolescence of other dams and hydropower installations will lead to a decrease in production and an increase in tariffs.

References

Chen L (2021) Environmental cost of sustainable development and climate change: can SAARC shift some liability with renewable energy and efficiency? Environ Sci Poll Res.https://doi.org/10.1007/s11356-021-15209-2

Gheorghe IF, Bodescu F, Martini M, Gheorghiu C, Abaza V, Diaconu DC, Bălteanu D, Dumitraşcu M, Gancz V, Gâştescu P, Ioja C, Minodora M, Maruşca T, Mateescu R, Mihalache M, Moldoveanu MM, Onete M, Pop OG, Radnea C, Silaghi D, Vânău G-O, Vintilă R, Virsta A (2017) MAES process in Romania. In: Nature for decision-making (N4D) English version. ISBN 978-606-8038-24-7, 89 pp

Güney T (2019) Renewable energy, non-renewable energy and sustainable development. Int J Sustain Dev World Ecol:1–9. https://doi.org/10.1080/13504509.2019.1595214

Güney T (2021) Renewable energy and sustainable development: Evidence from OECD countries. Environ Progr Sustain Energy 40(4). https://doi.org/10.1002/ep.13609

Gutierrez GM, Kelly S, Cousins JJ, Sneddon C (2019) What makes a megaproject? Environ Society 10(1):101–121. https://doi.org/10.3167/ares.2019.100107

https://ec.europa.eu/eurostat/statistics-explained/index.php?title=Electricity_price_statistics

https://www.hydropower.org/news/iha-releases-2019-hydropower-status-report-charting-growth-in-renewable-hydro

IEA (2019) Global energy & CO_2 status report 2019. IEA, Paris. https://www.iea.org/reports/global-energy-co2-status-report-2019

IHA (2019) Hydropower status report 2019. https://hydropower-assets.s3.eu-west-2.amazonaws.com/publications-docs/iha_2018_hydropower_status_report_4.pdf

IHA (2020) Hydropower status report 2020. https://www.hydropower.org/publications/2020-hydropower-status-report

Klimpt JÉ, Rivero C, Puranen H, Koch F (2002) Recommendations for sustainable hydroelectric development. Energy Policy 30(14):1305–1312. https://doi.org/10.1016/s0301-4215(02)00092-7

Manzano-Agugliaro F, Zapata-Sierra A, Alcayde A, Salmerón-Manzano E (2021) Worldwide research trends on hydropower. In: Jeguirim M (ed) Recent advances in renewable energy technologies. Academic Press, pp 249–280. https://doi.org/10.1016/B978-0-323-91093-4.00007-X

Strielkowski W, Civín L, Tarkhanova E, Tvaronaviˇciene M, Petrenko Y (2021) Renewable energy in the sustainable development of electrical power sector: a review. Energies 14:8240. https://doi.org/10.3390/en14248240

TREIA (2015). Definition of renewable energy. http://www.treia.org/renewable-energy-defined. Accessed 18 Jan 2022

Chapter 5
The Impact of Drought on the Hydropower Domain

Abstract Hydropower is based on the exploitation of water resources. Due to its circuit in nature, water thus ensures a continuous supply of hydropower plants. Significant variations in electricity production are due either to a high flow rate of liquid leakage or due to its drastic reduction. If the water flows are high and the electricity production is increased, all parties have something to gain. The producer delivers larger quantities and the consumer pays less. At the opposite pole are the diminished water resources due to drought. Both droughts and floods are normal features of a river's natural hydrological regime. In this context, the correct assessment of the situation in the case of reduced water flows helps to avoid the impossibility of honoring energy supply contracts.

Keywords Climate change · Drought · Indicators

As a multi-form phenomenon, drought can be considered an overall natural process and varies in time and space. There are several forms of description. Scientists usually investigate the main aspects of the drought with meteorological, agricultural and hydrological impacts separately, with the partial assessment of social and environmental influence generally taken into account by water managers. On the other hand, the concept of the phenomenon's impact on climate change and its relationship with climatic factors and geophysical conditions concerning water droughts and their scarcity and the link with reference periods are essential. Thus, examining the drought and the minimum run-off as a significant water drought alarm signal should also consider the external impact on the natural hydrological cycle.

As regards the hydrological cycle, it is observed that water constantly changes states, from atmospheric water (rain, snow, etc.) to surface water or groundwater, forming river beds and watercourses in river beds, moving from basin to basin or into the planetary sea before it becoming atmospheric water again.

The process is very complicated and depends on many physical processes, such as:

- precipitation, infiltration, run-off (surface or underground), evaporation, condensation, etc.
- temperature changes and energy transfer;

- physical and geographical conditions of the region;
- the particularities of the local climate;
- morphometric and morphological factors;
- land cover, urbanization and other factors such as hydrogeological and hydraulic conditions.

The water in the atmosphere moves with the air currents before falling in rainfall like burrow, rain, snow, sleet, hail, etc., on the ocean surface or the ground, causing surface and underground run-off. Part of the surface run-off forms watercourses (permanent and/or temporary) that move to the planet's ocean.

Part of the surface run-off on the verses is infiltrated close to the surface (hypodermic layer), generating soil moisture or deeper feeding groundwater or storing in lakes, anthropogenic accumulation or wetlands. Some waters seep deep into the ground, moving through paths of different pour lengths before reaching the unloading area or storage area of fresh water in aquifers for long periods. Groundwater moves slowly from surface run-off. Therefore, it is necessary to differentiate both physical processes to assess and investigate their impact, particularly in the case of their quantitative understanding and description. Part of the undersurface run-off (hypodermic) plays an essential role in water movement and loading of surface water, especially in conditions of minimal drought and run-off. Much of the water in the groundwater layer is returned to the earth's surface by dumping it into the rivers. After a certain period, they arrive in the planetary ocean or directly evaporate the atmosphere. Simultaneously, evaporation, sweat and purification occur, allowing water condensation and storage in the atmosphere.

As a phase of the hydrograph of the leak, the minimum run-off can be considered a characteristic of surface water. Various factors are causing it to emerge.

Most of them are linked and follow the water cycle, but the central part of both processes relates to local conditions and particularities. The main driver of the minimum run-off formation is the scarcity of precipitation, together with factors such as high air temperature, low relative humidity, high sun, low cloud coverage and radiation flow of available energy on the surface, so a high level of evaporation, sweat and water scarcity in the ground. This reduces the drain into the whiteness and entry of small flow rates into lakes, reservoirs and ponds, reducing wetlands and low levels of groundwater supply. This is closely linked to factors such as the physical and geographical conditions of the region, the particularities of the local climate, morphometric and morphological characteristics, soil cover, urbanization, and hydrogeological and hydraulic conditions, which determines the development of processes over time and space. Also taking into account the delays of some process phases and their differentiation by region. It is thus recognized that a given month can be regarded as drought by climatologists and farmers, who are mainly interested in precipitation falls, unlike hydrologists who are primarily interested in debits and may not record water drought but, for example, extreme humidity conditions as a result of snow melting.

5.1 Minimum Run-Off Phases

The minimum drain is characterized as one of the extremes of the hydrological regime. At the same time, it can be considered a result or indicator of drought. In practice, it is of particular interest in areas where there are no accumulation lakes or where the influence of run-off is seasonal or depends on human activity, and the flow regime is widely distributed. In this case, its essential understanding and determination are crucial. McMahon and Arenas (1982) stated that the minimum run-off is defined seasonally and directly linked to the annual solar cycle or its local or regional climatic effects (McMahon and Diaz Arenas 1982). The minimum run-off can also be absolute or relative. Different geographic areas are characterized by other behavior of minimal run-off. For the lower Danube basin with moderate continental climate conditions, the spill is most often due to rain-snow water, and the appearance of two dry seasons is representative. These seasons are summer and winter, the first season being more severe in terms of drought than the second. The seasonality and the severity of the minimum run-off depend on the climatic and physical geography of the basin. It may vary at different time scales, with spicy summers per river section cases. Considering the definition of drought as a period of abnormally dry weather, various parameters of minimum run-off can be observed.

The drought occurs for specific periods. There are six types of water-drought degrees (Beran and Rodier 1985) including:

- Run-off of between three weeks and three months, with drought during the regeneration and growing of plants, catastrophic for agriculture-dependent on irrigation directly from rivers without accumulation lakes;
- A minimum flow of significant or longer than the normal minimum required during the growing season of the plants. The germination period is not affected by this type of drought and has low consequences for agriculture;
- A significant deficit of total annual run-off. This affects hydroproduction and irrigation in large reservoirs;
- An annual maximum water level below the average river level. This may introduce the need for pumps for irrigation. This type of drought is linked to the yearly run-off deficit;
- Like the "drought" in northern Brazil, the drought lasts several years. The flow rates remain below the minimum threshold, or the rivers completely dry and stay dry for an extended period;
- Significant natural depletion of aquifers. This is difficult to estimate, as the identification of the actual level of aquifers is affected by the intensive use of groundwater during drought.

According to McMahon and Arenas (1982), the following concepts must be considered (McMahon and Diaz Arenas 1982). The minimum run-off period is usually defined as follows:

- Its duration is often the same as that of the dry season. This occurs when the lack or insufficient precipitation characterizes the season;

- If the absolute or lowest minimum flow rate is almost always equal to the daily minimum flow rate during one year;
- This series of minimum flows expresses the correspondence between fixed durations (expressed in several days) and flows that have not been exceeded during several consecutive days. For example:

  - Water flow not exceeded for 7–10 days;
  - Water flow not surpassed for 15 days,
  - Water flow not exceeded for one month.

The following cases may affect the process of determining the minimum flow rate—if the water is frozen or if the reserves feeding the riverbeds have been completed or are insufficient to generate surface run-off (although the underground flow continues). Thus, understanding and evaluating both processes and their characteristics are essential. Best estimate is subject to volatility adjustment—total (life other than health insurance, including non-life). This is of great importance in terms of the quantitative and qualitative aspects of water resources that are more sensitive during minimum run-off or droughts and in maintaining the ecological health of aquatic ecosystems to ensure a minimum acceptable flow rate in rivers. For example, there is a strong link between water quantity and quality. In the case of minimal run-off, the concentration of substances is higher than average run-off or flash floods. Thus, actual measurements and monitoring must be followed to achieve the actual results. Various technical measures may be used, possibly different from those used for average and maximum run-off measurements. Still, the most important aspect is their choice according to the specific nature of the area under investigation and the hydrological conditions.

Different methods could be used to practically implement monitoring and direct measurement of minimum run-off. Most of them are presented in the review of the World Metrology Organization, No 1044 (2010), where the organization of the network for minimum run-off rate studies for water resource development and adequate monitoring of droughts and minimum run-off is well presented (World Metrology Organization 2010).

5.2 Minimum Run-Off Indices

5.2.1 Basic Indices of the Minimum Run-Off

The minimum run-off can be analyzed in different ways of interpreting the time series of daily flows to produce summary information describing the minimum run-off regime of a river. The term "minimum run-off indices" is used for specific values derived from a minimum flow analysis. Some are unique values such as the recession constant, the base rate index or the average of a time series. These are called minimum run-off rates. More complex methods estimate the probability of minimal run-off.

For example, daily flows' cumulative frequency distribution (flow duration curve) describes the relationship between flow and time in percent when a given flow rate is exceeded. The theory of extreme values is used to estimate the probability of annual minima not being exceeded. What is essential in the two techniques is that the run-off duration curve considers all days in a chronological series and thus the duration in the percentage of the entire observation period in which a flow rate is exceeded.

In contrast, the theory of extreme values applied to the annual minimum series estimates the probability of non-exceedance in years or the average interval in years (recovery period) when the yearly minima are below a given value. It is, therefore, often helpful to specify the water resources management plan (Demuth and Kulls 1997a). Many water resource management decisions are based on these indicators:

The average flow (cubic meters/s)—is one of the most frequently used statistics in hydrology and water resource planning. It can be estimated from a chronological series of measurement data by summing all daily flows and dividing by the number of days with observations. It is usually calculated for a calendar year or for the hydrological year of the time series. It may also be estimated for specific months or seasons.

The insurance flow 95% $Q_{95\%}$ (cubic meters/s) is one of the most commonly used minimum run-off indicators operationally. It is defined as the flow rate exceeding 95 percent of the time interval.

It can be determined by ranking all debits (daily, monthly, yearly) and finding the flow rate with 95% of all values of the analyzed data line. This percentile and others ($Q_{90\%}$, $Q_{70\%}$, etc.) can be determined from the run-off flow path.

The minimum annual average flow rate of n-days: MAM (n-days). The minimum annual flow rate in n days (AM (n-days)) shall be the lowest average flow rate in n consecutive days in a year. The averaging intervals typically used, i.e., the values of n, are 1, 7, 10, 30, and 90 days.

AM (n-day) can be easily calculated by applying an n-day sliding average filter to a daily flow series and subsequently selecting a minimum of the filtered sequence. Thus, the average annual flow rate over the n-day interval (MAM (n-days)) is the average series of minimum yearly rates in n days.

Unlike percentiles in the duration curve, MAM (n-day) involves an aspect of the duration, included in the averaging interval. The minimum annual flow rates may be used to determine the distribution function to estimate the frequency or return period of the minimum flow rates. The 7-day average minimum yearly flow rate in a temperate climate is numerically similar to $Q95\%$ for most flow series.

Base run-off Index—BFI is the ratio of the base run-off to the total run-off calculated by a hydrograph separation procedure. It has been developed to characterize the hydrological response of the river basin according to the nature of the soil and geological conditions. BFI index values range from 0.15 to 0.20 for impermeable, flash floods-vulnerable basins to over 0.95 for permeable basins with high water storage capacity and stable flow regime.

Recession constant—is the C parameter of the recession curve.

5.2.2 Average Flows with Different Frequencies—Calculation Example

Determining the average flows and the different probabilities of excess is important for the economic (design of hydropower facilities, water supply for industry and agriculture), social (ensuring the need for water for domestic use) and ecological fields.

The statistical parameters of the average liquid flow value strings (Q—average flow, C_v—coefficient of variation and C_s—asymmetry coefficient) resulting from the classical statistical-mathematical processing calculations performed on the available data strings are not automatically the parameters of the analytical probability distribution curves.

In the example presented, to achieve this goal, the average annual flows for 10 sections located on the Argeş River, Romania were analyzed. The analyzed period covers from 30 to 65 years (1950–2014) depending on the data available for each section. Given that the data strings met the conditions of homogeneity and exceeded the limit number of min. 30 values per section, could be subjected to statistical-mathematical processing.

The main problem related to the construction of the empirical overinsurance probability curves is the assignment of an empirical overinsurance probability for each term of the ordered descending sequence. Defining the probability of overcoming-empirically securing the term with position i in the ordered sequence decreasing n terms by the ratio

$$\frac{i}{n} \times 100\% \tag{5.1}$$

is not entirely satisfactory, because the last value in the string would always have the probability of 100%, which cannot be true, because the emergence of new values is possible at any time. The theoretical formula of probability is:

$$p\% = \frac{i}{n+1} \times 100 \tag{5.2}$$

and was proposed by Weibull in 1939, and can be applied with very good results in the statistical-mathematical processing of average flows.

Other calculation formulas are also used for the probability of occurrence such as

$$p\% = \frac{2i-1}{2n} \times 100 \tag{5.3}$$

Hazen;

$$p\% = \frac{i-0.3}{n+0.4} \times 100 \tag{5.4}$$

Cegodaev;

$$p\% = \frac{3i - 1}{3n + 1} \times 100 \tag{5.5}$$

Tukey; which make $pn \neq 100\%$, and the higher the range of processed values, the higher the values of 99.99 (Rust et al. 2018).

In this example are presented the resulting insurance curves, for the 10 sections of the Argeş basin (Dârmăneşti, Ciumeşti, Voina, Mioveni, Bălileşti, Bahna Rusului, Malu Spart, Arefu, Nămăieşti, Amonte acumularea Vidraru) (Fig. 5.1).

The probabilities thus calculated, based on the equations (Eqs. 5.2–5.5), are written for each section on graphs specially constructed on a logarithmic and semi-logarithmic format. This ensures the insurance-overcoming curves. From these graphs, the values corresponding to the insurance flow 5%, 50% and 95% are extracted for each curve, with the help of which the standard deviation is calculated:

$$S = \frac{Q5\% + Q95\% - 2 * Q50\%}{Q5\% - Q95\%} \tag{5.6}$$

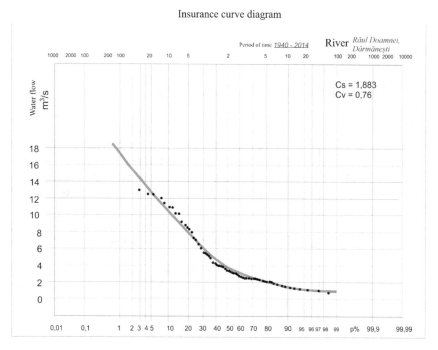

Fig. 5.1 Insurance curve exceeding the average annual flows at the hydrometric station Dârmăneşti, River: Râul Doamnei

Due to the fact that the statistical parameters of the string with a relatively small number of terms require corrections, the theoretical distribution curve of Pearson III probabilities was determined. With the help of the standard deviation and the Pearson III theoretical curve, the asymmetry coefficient C_s is determined, for each section by mathematical or direct interpolation. The same is done with the coefficients $\varphi 5\%$, $\varphi 50\%$ and $\varphi 95\%$. Then according to the equation

$$(\varphi p\% \times C_v + 1) \times Qo = Qp\%, \tag{5.7}$$

the values for the insurance flows $Q5\%$, $Q50\%$ and $Q95\%$ are calculated. The coefficient of variation (C_v) is calculated with the relation:

$$C_v = \frac{Q5\% - Q95\%}{Q95\% * \varphi 5\% - Q5 * \varphi 95\%}. \tag{5.8}$$

The results obtained from the calculation of the probabilities of exceeding the average annual flows corresponding to the insurance probabilities of 1%, 5%, 10%, 80%, 90% and 95% are presented in the table of synthesis of the parameters (Table 5.1).

5.3   Hydrological Drought Indices

5.3.1   Standardized Flow Rate Index (SFI)

Most drought indices (PDSI, CMI, SWSI, and SPI) are derived from meteorological observations (especially precipitation and sometimes air temperature). Droughts should be evaluated and monitored, where possible, using other types of data (e.g., liquid flow). McKee et al. (1993) have suggested that the precipitation procedure (for the SPI index) can be adapted to other variables such as liquid flow and groundwater flow (McKee et al. 1993).

The standardized flow rate index (SFI) was a simple and helpful tool in researching, monitoring and managing water droughts in a controlled river system (Wen et al. 2011).

This indicator (standardized flow rate index (SFI) calculates the standardized precipitation index (SPI) using the average monthly flow series in natural flow rates instead of the total monthly rainfall quantities.

Table 5.1 Hydrological parameters with various degrees of excess determined in the Argeş River basin

No.	River	Hydrometric station	Q_0 m³/s	C_s	C_v	C_v/C_s	Water flow, determined with different degrees of achievement m³/s					
							$Q_{1\%}$	$Q_{5\%}$	$Q_{10\%}$	$Q_{80\%}$	$Q_{90\%}$	$Q_{95\%}$
1	Arefu	Arefu	0.25	1.57	0.52	0.33	0.71	0.52	0.44	0.15	0.12	0.11
2	Argeş	Malu Spart	39.01	0.90	0.39	0.43	82.51	66.08	58.32	25.60	21.12	18.14
3	Argeşel	Mioveni	1.23	1.90	0.65	0.34	4.068	2.829	2.277	0.598	0.494	0.446
4	Bratia	Balileşti	3.54	1.50	0.52	0.35	10.00	7.39	6.19	2.10	1.72	1.51
5	Nămăeşti	Nămăieşti	0.54	1.167	0.30	0.26	1.24	1.01	0.90	0.48	0.43	0.40
6	Râul Doamnei	Bahna Rusului	3.92	2.20	0.86	0.39	15.70	10.32	7.89	1.34	1.04	0.91
7	Râul Doamnei	Ciumeşti	13.12	1.40	0.51	0.36	35.96	26.89	22.69	7.77	6.33	5.44
8	Râul Doamnei	Dărmăneşti	5.05	1.833	0.76	0.41	18.84	12.90	10.26	2.02	1.49	1.20
9	Târgului	Voina	2.04	0.40	0.27	0.68	3.48	3.00	2.77	1.57	1.36	1.20
10	Argeş	Vidraru	7.51	0.50	0.22	0.44	11.65	10.32	9.57	6.04	5.44	4.99

5.3.2 Standardized Groundwater Level Index (SGI)

A new index for the standardization of groundwater level time series and the characterization of associated droughts, the standardized groundwater level index (GI), is described by Bloomfield and Marchant (2013). The SGI shall be formed based on the standardized precipitation index (SPI) of differences in shape and particularities of the groundwater level series and precipitation. The SGI shall be estimated using normal non-parametric tests transforming the groundwater level series for each calendar month. These monthly estimates are combined to form the continuous series of the index. The SGI has been calculated for 14 to 103 years of groundwater level hydrographs in various aquifers compared to SPI for the same areas. The link between SGI and SPI depends on the nature of the study area and the periods for which the SPI is calculated, leading to the strongest link between SGI, SPI and q_{max}, which varies from one area to another (Loon et al. 2017; Rust et al. 2018; Hellwig et al. 2020). However, there is a consistent linear positive correlation between a significant size of the autoclave range in the SGI, m_{max} and q_{max} index series in all areas concerned. Given this correlation between SGI m_{max} and SPI q_{max} and those periods with minimum SGS values can coincide with previously independently documented droughts, SGI is considered a robust and important index of hydrogeological drought (Balacco et al. 2022).

The maximum time of hydrogeological droughts defined by the SGI is an increasing function of m_{max}, meaning that relatively prolonged hydrogeological droughts are generally more prevalent in places where GGI has a reasonably large self-referencing range. Based on the correlations between m_{max}, the density of the medium unsaturated area and the hydraulic diffusion capacity of the aquifer layer, the source of the autoclaving in the SGI is declared dependent on the dominant aquifer passage the special accumulation features.

5.3.3 Standardized Run-Off Index (SRI)

The standardized run-off index (SRI) is an appropriate tool in identifying and characterizing the water drought concerning the drained water flow. A standardized flow rate assessment ensures that its abnormal conditions are counted and graded. It has a common basis with the standardized precipitation index (SPI), but the flow specificity is used to calculate the index in the evaluation process. It was developed by McKee et al. (1993) and recommended for the hydrological drought characterization by referring to the estimation procedure by Shukla and Wood (2008) (McKee et al. 1993).

In-depth investigations have been carried out into the possibility of practical implementation in the EU Member States. They have been conducted in the context of the EU drought and water scarcity Expert Group (Schmidt and Benítez-Sanz 2012). The results have been compared with historical data and other drought indicators, such as SPIs or other local indices, concluding that:

- The indicator accurately represents the results;
- SRI can identify past as well as future droughts;
- The procedure for estimating the SRI index is easy to apply;
- The appropriate time scale depends on the type of basin and the objectives of the evaluation.

For the objectives of the phenomenon's identification, its categorization may be associated with the severity limits presented (Keyantash and Dracup 2002):

Categories	Values SRI
Extremely wet	SRI > 1.65
Very wet	1.65 < SRI < 1.28
Moderate	1.28 < SRI < 0.84
Almost normal	0.84 < SRI < − 0.84
Moderate drought	− 0.84 < SRI < − 1.28
Severe drought	− 1.28 < SRI < − 1.65
Extreme drought	SRI < − 1.65

5.3.4 Minimum Run-Off Indices for Drought Fault Finding

The minimum run-off indices aimed to assess the minimum run-off and diagnose water droughts. According to McMahon and Arenas (1982), Beran and Rodier (1985), Dakova (2004), these indices are used to characterize the minimum run-off or as a measure of the severity of the drought in a year (McMahon and Diaz Arenas 1982; Beran and Rodier 1985; Dakova 2004). The following main characteristics are necessary to define the minimum run-off:

- The average of the minimum flow rate for n consecutive days;
- Data on their occurrence;
- The frequency assigned to the phenomenon.

The magnitude of the minimum run-off is the quantity of water flowing through a given section of a white for a specific period, thus determining the amount of water available for use. Duration depends on natural conditions and man-made effects and may reflect particular water use methods (e.g., irrigation cycles). The duration also depends on a user-tolerant water scarcity period or other requirements. This determines the start and end of the process, while the frequency feature requires a timescale investigation of the event's occurrence.

Obtaining the necessary information is related to using the long-term site data or analog to a comparative record. Different approaches for estimating them may be applicable, such as:

- simple ranking;
- an empirical probability of non-exceedance based on series values;
- a deeper approach based on the choice of statistical distribution and the layout on the graph.
- alternative frequency description methods that express the probability of the recovery period in percent is being the average range of occurrence of selected non-exceeded flows.

As Dakova (2004) summarizes about the events of the year or the annual variability, drought could be considered as a "short-term" drought or as a "long-term" drought, depending on the temporal scale implemented. For the analysis of the first category, data and daily drought threshold values based on percentage points of the flow duration curve are used in several works (Dakova 2004). For most regions, values—threshold with the probability of 70 and 90% are considered appropriate. However, both values are zero (Servat and Demuth 2006). In another approach, use percentages of the duration curve of daily flows with probabilities of exceedance of more than 30 % defined as minimum flow by Demuth and Kulls (1997b). According to Vlach et al (2020), the daily average flow rate below a specific threshold value, regarding the liquid flow with a 95% probability, is considered the lowest minimum flow rate (Vlach et al. 2020). The same author has accepted that long-term droughts are connected with different physical processes during the dry season or for a more extended period. Monthly or annual data are recommended for long-term drought analysis.

In the same work, the author describes indirect methods of assessing drought using different mathematical approaches. One of the most applied criteria is the cumulative value criterion explained by Herbst (1966), Srinivasan and Philippose (1993) with the starting point and termination of the determination by comparing the cumulative values of the hydrological variable over an appropriate time period (Herbst 1966; Srinivasan and Philipose 1993). It is also noted that this method is suitable to identify the drought in a series of monthly times. Such a method usually aims to identify drought for agricultural purposes. It, therefore, includes a preliminary assessment of actual rainfall and the sum of recent rain and weighted deviation from previous rainfall with the weighting describing the transfer effect. There are also some assumptions: short periods with limited deficits are typical for most climates. The ending of drought does not necessarily coincide with a surplus range. Such surplus may not be appropriate to be recovered by drought conditions. On this basis, different tests have been performed to determine the beginning and end of drought by comparing the cumulative values of the hydrological variables as a function monthly.

Correa intended to identify the beginning of the drought with the criterion of a single value, while partial or total recoveries from previous events were taken into account for the terminal stage (Correa et al. 2016). They assumed that the drought begins with the first interval below a specific threshold value called the critical level, usually lower than the average. The end of the drought does not coincide with the end of the negative run since a certain amount of surplus must be accumulated to recover the drought conditions. Such accumulated surplus is related to the highest threshold value (usually average) accepted as the recovery level. It must be equal to a fixed percentage of the previously accumulated deficit to determine the end of the drought. Thus, the identification of the drought duration includes two phases: One describing the deficit conditions and the other representing the recovery time after the drought effect by objective deficiencies according to the required definition of the critical and recovery levels. According to Rodrigues et al. (1993), the critical level equal to the 20 percentiles of the hydrological variables can be applied, the recovery level being considered equivalent to the mean and a percentage of the recovery volume equal to 40% of the previous deficit is appropriate for the determination of the end of a drought (Rodrigues et al. 1993; Grosser and Schmalz 2021). The selection of proper levels of the threshold value is a little arbitrary. The deficit rate is probably the most complicated for the assessment.

According to the authors, there is no apparent difference between the minimum run-off and the identification of the drought.

A further approach to the minimum run-off was presented by McMahon and Diaz Arenas (1982). The minimum run-off data can be considered a series of hydrological data from different variables. In general, the time series may consist of both random and non-random elements.

Non-random items in the series may appear as one or more of the following components:

- a trend;
- a random movement around the trend;
- a seasonal movement;
- a determinant component (generally measured by the correlation coefficient of the series).

The general idea of the trend is that of a smooth movement of a time series that extends over a long period of time. In hydrological time series, a trend detected in a given sequence may be due to:

- a slow and continuous variation of meteorological conditions (climate change) or a long periodic cyclical variation of the climate. (The series represents an increasing or decreasing branch of a cycle);
- a change in the physical and geographical conditions of the basin due in particular to human activity.

The first point is complex hydrological analysis, found in references such as Diaconu and Serban (1994). The second type of modification is permanent as long as human activity continues at the same level. This destroys the homogeneity of the data series and must be considered in the data evaluation.

Distinct approaches are generally applied even for the characterization of water droughts. It is processed in a catchment or reception basin in hydrological terms, but the following results can be widely generalized concerning investigation needs as set out below.

5.4   Statistical Methods and Approaches for Assessing the Minimum Run-Off

The minimum run-off regime can be analyzed in various ways, depending on the initially valid data and the type of output required information. Therefore, there is a wide range of indices and minimum flow measurements. The term 'minimum run-off rate' refers to the different methods developed to analyze, often in graphic form, the minimum run-off rate of a river. The term "minimum run-off rate" is often used to define particular values obtained from any minimum run-off measurement (sometimes it is difficult to separate them from each other) (Smakhtin 2001).

The statistical analysis of the minimum run-off indicates the availability of water in whites when the water requirements are likely to be out of reach. For this reason, minimum run-off methodologies are required by all fora in charge of water resource management for water planning, management and regulatory activities. These activities include:

- the development of environmentally effective river basin management plans,
- support for and permit new water extraction, inter-basal transfers and wastewater flows, determination of threshold values for minimum flow rates for the maintenance of the aquatic biosphere and planning and regulation of land use.

Commercial, industrial and hydroelectric entities require minimum run-off methodologies to determine the availability of water for water supply, wastewater flow and energy production (Ries and Friesz 2007).

Regarding international studies on the minimum run-off and drought, Tallaksen and van Lanen (2004) claim extensive literature on the various processes that operate during the minimum run-off or a period of drought. In particular, the basin response and the minimum run-off regime of a river or region have been described. Less material is available about methods for assessing minimum run-off and drought, including prediction, forecast and estimation in unmonitored sections. The latter has already been stipulated in research relevant to the forecast of the minimum run-off for the hydrometer-controlled sections and is still valid (Tallaksen and Lanen 2004).

Smakhtin (2001) concluded that in the river basin's integrated and environmentally sustainable management, the minimum flows could be seen as a dynamic concept rather than described by only one characteristic of the minimum run-off (Smakhtin 2001). Therefore, priority is the flow-rate time series from which the variety of minimum run-off indices can be extracted to meet different management and engineering purposes.

The most common regionalization analysis of run-off is the methods for the duration of the leak, including, among other things, the base run-off index, the leak rate analysis, and the recessionary parameters (Demuth and Kulls 1997a).

The "Manual on minimum run-off estimation and forecasting", published by the World Meteorological Organization, provides a comprehensive summary of how to analyze the series of liquid flows, especially the minimum flow rates (World Metrology Organization 2010). Although the indices and methods for calculating the minimum run-off have been well documented in this manual, comprehensive software is missing, which provides a quick and standardized calculation of the minimum run-off statistical analysis. This software is based on the open-source R statistical program. It extends to studying the daily average flow time series, allowing a fast and standardized calculation of the minimum run-off analysis (https://www.r-project.org/) (Cammalleri et al. 2017).

There are different analyses of the time series of daily average flows to produce concise information describing the minimum drainage regime of a river. This application package provides functions to calculate the defined statistics and has graphs similar to those in the manual.

In Romania, researchers from the National Institute of Hydrology and Water Management (NIHWM) collaborated with the European FRIEND working Group—minimum water run-off and drought on the topic "minimum run-off indices," to prepare a significant publication about the current minimum run-off conditions in Europe. The study calculated the run-off of minimum indices based on the application package mentioned above (Diaconu and Serban 1994; Diaconu 2005, 2016, 2018; Cojoc 2016).

Based on the results obtained, as a result of this package of applications, indices of minimal run-off have also been calculated. These statistical methods and approaches can undoubtedly be a comprehensive and essential analysis in assessing the minimum run-off.

5.5 Conclusion

Determining the variability of river run-off is currently possible due to the background of hydrological data recorded over a long period of time. These databases allow the application of specific indices to determine the exceptional values that can be recorded with a certain frequency. This chapter aims to briefly present some of the most often used indices used in determining the distribution of water resources.

The planning of electricity production, which uses water resources directly, must be done in close connection with the anticipation of changes in local climate and hydrological regime. This helps in avoiding critical situations when the water tank is depleted and can no longer provide the volumes of water needed for main or related uses thus making it difficult for the electricity producer to honor their contracts.

References

Balacco G, Alfio MR, Fidelibus MD (2022) Groundwater drought analysis under data scarcity: the case of the Salento Aquifer (Italy). Sustainability 2022(14):707. https://doi.org/10.3390/su1402 0707

Beran MA, Rodier JA (1985) Hydrological aspect of drought. UNESCO—WMO, Geneva

Bloomfield JP, Marchant BP (2013) Analysis of groundwater drought building in the standardized precipitation index approach. Hydrol Earth Syst Sci 17:4769–4787. https://doi.org/10.5194/hess-17-4769-2013

Cammalleri C, Vogt J, Salamon P (2017) Development of an operational low-flow index for hydrological drought monitoring over Europe. Hydrol Sci J 62(3):346–358. https://doi.org/10.1080/02626667.2016.1240869

Cojoc GM (2016) Analiza regimului hidrologic a râului Bistrița în contextul amenajărilor hidrotehnice. Terra Nostra, Iași. ISBN 978-606-623-061-2

Correa A, Nixon DM, Russel C (2016) An Economic and nutritional evaluation of pricklypear as an emergency forage supplement. Tex J Agric Nat Resour 1:41–44

Dakova S (2004) Low flow and drought spatial analysis. Paper presented at the BALWOIS 2004, conference on water observation and information system for decision support, Ohrid, Republic of Macedonia, 25–29 May 2004

Demuth S, Kulls C (1997a) Probability analysis and regional aspects of drought in Southern Germany. Hydrological Journal 240:97–103

Demuth S, Kulls C (1997b) Probability analysis and regional aspects of drought in Southern Germany. In: Sustainability of water resources under increasing uncertainty (proceedings of Rabat Symposium S1, April 1997b). IAHS Publ. No. 240, pp 97–103

Diaconu DC (2005) The water resources in the Buzau river basin, Universitară, Publishing House Bucharest, 238p. ISBN 973-7787-57-9

Diaconu DC (eds) (2016) Spatio-temporal analysis of water resources in the Argeș Basin. Transversal Publishing House, Targoviste. ISBN 978-606-605-140-8

Diaconu DC (2018) Water from a geographic perspective. Transversal Publishing House, Târgoviște, 161 p. ISBN 978-606-605-183-5

Diaconu C, Serban P (1994) Sinteze si regionalizari hidrologice. Editura Tehnica, Bucuresti, 388 p. ISBN 973-310-6380

Grosser PF, Schmalz B (2021) Low flow and drought in a german low mountain range basin. Water 13:316. https://doi.org/10.3390/w13030316

Hellwig J, de Graaf IEM, Weiler M, Stahl K (2020) Large-scale assessment of delayed groundwater responses to drought. Water Resour Res 56:e2019WR025441. https://doi.org/10.1029/2019WR 025441

Herbst PH (1966) A technique for the evaluation of drought from rainfall data. J Hydrol 4:264–272 https://www.r-project.org/

Keyantash J, Dracup JA (2002) The quantification of drought: an evaluation of drought indices. Bull Am Meteor Soc 83:1167–1180

Loon AFV, Kumar R, Mishra V (2017) Testing the use of standardised indices and GRACE satellite data to estimate the European 2015 groundwater drought in near-real time. Hydrol Earth Syst Sci 21:1947–1971. https://doi.org/10.5194/hess-21-1947-2017

McKee TB, Doesken NJ, Kleist J (1993) The relationship of drought frequency and duration to time scales. Paper presented at 8th conference an applied climatology, Anaheim, Calif., Amer. Meteor. Soc., pp 70–184

McMahon TA, Diaz Arenas A (1982) Methods of computation of low streamflow, A contribution to the International Hydrological Programme. UNESCO, Mayenne, Imprimerie de la Manutention

Ries KG, Friesz PJ (2007) Methods for estimating low-flow statistics for Massachusetts streams. USGS Water Resources Investigations Report 00-4135. US Geological Survey, Washington D.C

Rodrigues R, Santos MA, Correia FN (1993) Appropriate time resolution for stochastic drought analysis. In: Marco JB, Harboe R, Salas JD (eds) Stochastic hydrology and its use in water resources systems simulation and optimization. NATO ASI Series (Series E: Applied Sciences), vol 237. Springer, Dordrecht. https://doi.org/10.1007/978-94-011-1697-8_15

Rust W, Holman I, Corstanje R, Bloomfield J, Cuthbert M (2018) A conceptual model for climatic teleconnection signal control on groundwater variability in Europe. Earth Sci Rev 177:164–174. https://doi.org/10.1016/j.earscirev.2017.09.017

Schmidt G, Benítez-Sanz C (2012) Topic report on: assessment of water scarcity and drought aspects in a selection of European Union river basin management plans. Study by Intecsa-Inarsa for the European Commission (under contract "Support to the implementation of the Water Framework Directive (2000/60/EC)" (070307/2011/600310/SER/D.2))

Servat E, Demuth S (eds) (2006) FRIEND—a global perspective 2002–2006, IHP Non-serial publications in hydrology. German IHP/HWRP Secretariat, Koblenz

Shukla S, Wood AW (2008) Use of a standardized run-off index for characterizing hydrologic drought. Geophys Res Lett 35:L02405. https://doi.org/10.1029/2007GL032487

Smakhtin VU (2001) Low flow hydrology: a review. J Hydrol 240(3–4):147–186. https://doi.org/10.1016/S0022-1694(00)00340-1

Srinivasan K, Philipose MC (1993) Stochastic modeling for drought analysis. In: Engineering hydrology. ASCE, pp 323–328

Tallaksen LM, van Lanen HA (eds) (2004) Hydrological drought—processes and estimation methods for streamflow and groundwater. Dev Water Sci 48

Vlach V, Ledvinka O, Matouskova M (2020) Changing low flow and streamflow drought seasonality in Central European headwaters. Water 12(12):3575. https://doi.org/10.3390/w12123575

Wen L, Rogers K, Ling J, Saintilan N (2011) The impacts of river regulation and water diversion on the hydrological drought characteristics in the lower Murrumbidgee River, Australia. J Hydrol 405(3–4):382–391

World Metrology Organization, No 1029 (2008) Operational hydrology report (OHR) No. 50. ISBN: 978-92-63-11029-9

World Metrology Organization, No 1044 (2010)

Chapter 6
Force Majeure

Abstract This chapter presents the significance of the phrase force majeure, its formulation and application in commercial contracts when the impossibility of providing the services assumed by said contract arises due to unforeseen causes independent of the will and foresight of the signatory parties.

Keywords Concept · Law · Contracts · Clause

6.1 Understanding the Concept of Force Majeure

Force majeure is a French term that means "greater force". It is related to the concept of an act of God, an event for which no party to the contract can be held liable. However, force majeure also includes human actions, such as armed conflicts. In general, for events to constitute a force majeure, they must be unpredictable, external to the parties to the contract, and unavoidable. These concepts are defined and applied differently depending on the jurisdiction.

The concept of force majeure has its origin under French civil law and is a standard accepted in many jurisdictions deriving their legal systems from the Napoleonian Code. In standard law systems such as those in the United States and the United Kingdom, force majeure clauses are acceptable but should be more explicit about the events that would trigger the clause (https://us.eversheds-sutherland.com/portal resource/USForceMajeureGuideUS.pdf).

What is a force majeure clause?

A "force majeure" clause is a contractual provision that exempts the parties from fulfilling their contractual obligations when circumstances arise beyond their control, making the activity commercially impracticable, illegal or impossible. In the absence of a force majeure clause, the parties to a contract are left to the narrow doctrines of the contract under ordinary law on "impracticability" and 'frustrating the purpose,' which rarely lead to an excuse for performance. Instead of being based on common law, contract signatories can gain more flexibility in times of crisis through a carefully negotiated force majeure clause. Whether negotiating with or without the assistance of a legal adviser, the following critical elements of a force majeure clause should be addressed (Article 1218 of the French Civil Code; General Contract Clauses).

Force majeure clauses are contractual clauses that alter the obligations and/or liabilities of the parties under a contract where an extraordinary event or circumstance which is not under their control prevents one or all of the signatories of the contract from fulfilling their obligations.

Depending on how they are drafted, such clauses may have some consequences, including the exclusion of the affected party from performing the contract in whole or in part; the exemption of that party from penalties, giving them the right to suspend or request an extension of the time for the performance of the contract; or giving that party the right to terminate the contract.

Deepak Dayal, Managing Partner at Dayal Legal Associate, states that in English and Scottish law, force majeure refers to the contract, not general common law (https://www.linkedin.com/pulse/force-majeure-deepak-dayal-1c/) (Treitel 2014a). It is therefore different from other legal systems where force majeure is a general legal concept and where courts can declare that a particular event, such as a pandemic such as COVID-19, is an event of force majeure. Such an event may lead to the cancellation or postponement of the performance of the contract concluded by the parties (Coronavirus COVID 19).

As a result, if a particular term exempts a party from contractual liability under English and Scottish law, it will depend on the exact wording used in the term, on the allocation of risk between the parties to the contract as a whole, on the circumstances in which the parties have entered into the contract and on the situation that has arisen.

Anticipate and specify force majeure events

It is essential to establish which circumstances will be covered by the force majeure clause. Provisions often cover natural disasters such as hurricanes, floods, earthquakes, and extreme weather events, sometimes referred to as "Acts of God". Other covered events may include wars, acts of terrorism or terrorist threats, civil wars, strikes or interruptions to work, fires, diseases, epidemics, medical outbreaks, and reduction of transport facilities that prevent or delay the performance of at least twenty-five percent of the contract (https://www.law.cornell.edu/wex/force_majeure) (Ronen 2010; Milovanovic and Dumic 2021; O'Reilly 2021; Hennings et al. 2022).

Courts tend to interpret force majeure clauses narrowly; only the listed events and events similar to those listed will be covered. For example, while terrorist acts may be a specified force majeure event, it does not necessarily result that a court would also excuse the performance of a party based on terrorist "threats". It is essential to specify any circumstances that you foresee might prevent the meeting from running.

Whenever possible, consider the location of the event and any special needs or responsibilities of your company and contract participants. What types of weather incidents are likely to occur in the project area? If there are major disruptions to transportation systems due to any foreseeable reason, will your employees be prevented from working? What percentage of work would need to be impossible to perform in order for force majeure to be considered valid? The question and answer to these types of questions will help you anticipate and specify the most critical force majeure events for your business. Even so, not all potential events can be specified or anticipated in the contract. A final list should be added to the contract, such as "and any other events, including emergency or non-emergency situations", to cover other unforeseeable events (https://www.law.cornell.edu/wex/force_majeure).

Avoid restrictive terminology

It is only reasonable to find the wording of force majeure important when performing contracts that limit the excuse of the parties' performance obligations only when it would be "impossible" to do so because of unexpected circumstances. Impossibility is a high threshold; many circumstances will make it impossible to legally fulfill an activity, even if it would still be physically possible to finish some or even all of it. For greater flexibility, consider the excuse for lack of performance instead when it would be "inadmissible, commercially impracticable, illegal or impossible".

In addition, even if you negotiated a specified list of force majeure events, make sure you read the punctuation signs that appear in the list carefully. A comma can significantly change the scope of the force majeure clause. For example, adding the words "or any other emergency beyond the control of the parties" to the end of a list of specified force majeure events serves to restrict the purpose of the occurrence only to "emergencies". With such language, non-emergency circumstances will not be covered (https://us.eversheds-sutherland.com/portalresource/USForceMajeureG uideUS.pdf).

Consider the drop in performance due to force majeure

Although a force majeure clause should always allow for the complete cancellation of a contract without penalty, the cancellation will not always be the preferred course of action by the contracting parties. There may be circumstances in which continued contract performance may be preferred, even though the force majeure event will likely lead to a longer delay than expected. The force majeure clause must be drafted to pardon the liability associated not only with the non-performance (i.e., cancellation of the contract) but also with the non-performance (i.e., failure to comply with minimum guarantees) (https://www.jdsupra.com/legalnews/the-coronavirus-and-force-majeure-56194/, https://www.lawgazette.co.uk/legal-updates/force-majeure-in-2020/5103626.article).

A carefully negotiated force majeure clause is an essential tool for reducing the liability risk associated with canceling or reducing a scheduled meeting in response to a disaster. Where significant resources are available, signatories of the contract should consider requesting consultation of a legal adviser before signing contracts and should also consider obtaining insurance. Taking appropriate precautions from the outset can ensure that you will have the flexibility to make the best decision and overcome the major crises that may arise even under the most severe circumstances.

As the impact of the COVID-19 pandemic develops every day, supply chains are significantly disrupted in all economic and social systems. A persistent question is whether a force majeure clause excuses the parties from fulfilling their obligations or doing so on time.

If English or Scottish law is applied, the answer depends on the particular circumstances and the preparation of the relevant contractual arrangements. The position is similar, for example, in Singapore. Despite previous events, such as SARS and Ebola, there is no case law reported in these jurisdictions on applying force majeure clauses in the context of epidemics or pandemics (https://us.eversheds-sutherland.com/portal resource/USForceMajeureGuideUS.pdf, https://www.linkedin.com/pulse/force-maj eure-deepak-dayal-1c/).

Is COVID-19, for example, a force majeure event?

Force majeure clauses will generally adopt one of the following approaches for defining the type of event which, depending on its impact, may relieve a part of contractual liability:

Presentation of specific events. These may include events such as war, terrorism, earthquakes, hurricanes, droughts, government acts, or epidemics. Where the term epidemic or pandemic has been used, it will clearly cover the COVID-19 pandemic.

A government act is issued if a representative body has imposed travel, work, quarantine, trade embargo restrictions, closed buildings, localities, or borders. However, the position is less clear if the government makes recommendations rather than ordering through legal acts.

Where no relevant event is specifically mentioned, it is a question of interpretation of the clause if the parties intended to be covered by such an event. This implies that the list of events included was intended to be exhaustive or non-exhaustive. Unless specific words are used to suggest that a list is not exhaustive, it may be difficult to claim that parties who have established a list of specific events but have not included a particular event, such as an epidemic, intended—however, that event to be covered.

The establishment of general criteria. Contracts could, for example, relate to events or circumstances "beyond the reasonable control of the parties". Determining whether it covers the problems arising from the COVID-19 pandemic is a matter of interpretation and is highly dependent on the factual situation (https://us.eve rsheds-sutherland.com/portalresource/USForceMajeureGuideUS.pdf, https://www. linkedin.com/pulse/force-majeure-deepak-dayal-1c/).

In unprecedented circumstances, such as the presence of the COVID-19 pandemic, courts are likely to be generous in interpreting such forms when faced with parties who have encountered genuine difficulties in the performance of the contract. However, such parties will still have to demonstrate that their non-execution or late performance was out of their control and could not have been prevented or mitigated.

- A combination of the above clauses can provide a list of specific criteria, such as fire, floods, droughts, war and so on, along with a clause using broad, unspecific wording such as "or any other causes beyond our control". Although most of it will depend on the interpretation of the particular terms used, the general wording mentioned in the final, broad clause will usually be interpreted widely rather than being limited to only certain kinds events similar to those usually mentioned.

These are legislative situations where state-specific requirements for force majeure clauses are presented, such as New York, Florida, California, Texas, Illinois.

Under New York law, a vital issue in determining whether a party can successfully invoke a force majeure clause is whether the clause lists the specific event allegedly hindering performance (Phibro Energy, Inc. v. Empresa De Polimeros De Sines Sarl 1989). As mentioned above, some force majeure clauses list "epidemics" or "pandemics" as force majeure events, fapt prezentat si de Lawrence P. Rochefort and Rachel E. McRoskey in The Coronavirus and Force Majeure Clauses in Contracts (https://www.akerman.com/en/perspectives/the-coronavirus-and-force-majeure-clauses-in-contracts.html). Centers for Disease Control and Prevention defines an epidemic as an outbreak of disease that infects communities in one or more areas, and a pandemic is an epidemic that spreads around the globe. Suppose a contract in question lists epidemics or pandemics as a force majeure event. In that case, the complainant party may claim coronavirus qualifies because it has been officially declared a pandemic by the World Health Organization.

On the other hand, epidemics and pandemics are still recorded in the past, and it is not a situation that humanity has never experienced before (1918 pandemic (H1N1 virus); 1957-1958 pandemic (H2N2 virus); 1968 pandemic (H3N2 virus); 2009 H1N1 pandemic (H1N1pm09 virus)) (https://www.cdc.gov/flu/pandemic-resources/1918-pandemic-h1n1.html, https://www.cdc.gov/flu/pandemic-resources/1957-1958-pandemic.html, https://www.cdc.gov/flu/pandemic-resources/2009-h1n1-pandemic.html) (Jester et al. 2020).

Suppose a force majeure clause does not list the epidemic or pandemic as a trigger event. In that case, the coronavirus may be covered by an act issued by a government authority in certain areas, given that many governments (United States, Italy, Great Britain, etc.) have established blockings to prevent the spread of coronavirus.

However, if a force majeure event occurs, further analysis is required to determine whether the invocation will be successful. In New York, the event of force majeure must be unforeseen, and the party seeking to invoke the force majeure clause must try to fulfill its contractual duties despite the event (Rochester Gas & Elec. Corp. v. Delta Star, Inc. 2009). However, some jurisdictions, including Texas, do not require that the force majeure event be unpredictable (Perlman v. Pioneer Ltd. P'ship 1990).

Under Florida legislation, a party wishing to invoke a force majeure clause must show that the event of force majeure was unforeseeable and took place outside the party's control. This means that the complainant party must demonstrate that the event could not have been prevented or exceeded and, in addition, there can be no fault or negligence on the part of the complainant party (https://www.law.cornell.edu/wex/force_majeure).

In California, force majeure is not necessarily limited to the equivalent of an Act of God. However, the test is whether, in the particular circumstances, an event beyond acceptable foreseeability which could not have been prevented by exercising reasonable caution, diligence and care has taken place. Even in the case of force majeure mentioned in a contract, a mere increase in costs does not excuse performance unless there are extreme and unreasonable difficulties, expenses, injuries, and/or losses (https://us.eversheds-sutherland.com/portalresource/USForceMajeureGuideUS.pdf).

Under Texas law, unless expressly included in a contract, parties wishing to invoke a force majeure clause to excuse non-performance are not obliged to exercise reasonable diligence to perform or exceed the event of force majeure (Sun Operating Ltd. P'ship v. Holt 1998). However, if the parties have contracted for this, determining whether a party has exercised reasonable diligence is intensive. It should be assessed on a case-by-case basis (El Paso Field Servs., L.P. v. Mastec N. Am., Inc. 2012). "reasonable diligence" is defined in Texas legislation as "such diligence that a regular prudent and diligent person would exercise in similar circumstances" (El Paso Field Servs., L.P. v. Mastec N. Am., Inc. 2012; Levashova 2020).

Therefore, the Commission believes that the measure in question does not involve State aid. This debt is "linked to the debt of good faith [and] is read in all express contracts, unless waived" (Commonwealth Edison Co. v. Allied-Gen. Nuclear Servs. 1990).

Some contracts also require the claimant party to notify the other parties before invoking a force majeure clause. If the complaining party fails to give due notice as provided in the contract, a force majeure clause could be invoked successfully.

Companies wishing to invoke the force majeure clause in their contracts may have a strong argument that a coronavirus pandemic is an unforeseen event unless the parties have concluded the contract after its outbreak. Suppose businesses have also tried to perform their contractual duties despite the coronavirus outbreak, which is necessary even under a particular contract. In that case, these questions need to be assessed on a case-by-case basis (https://us.eversheds-sutherland.com/portalres ource/USForceMajeureGuideUS.pdf).

Certificates of force majeure

It is important to note that, due to the scale of the coronavirus outbreak and the government-imposed blockages in China, a quasi-governmental agency called China's Council for the Promotion of International Trade (C.C.P.I.T.), supported by the Beijing Ministry of Commerce, has provided companies in China with force majeure "certificates". C.C.P.I.T. issues certificates of force majeure where undertakings can provide documents showing that they cannot fulfill their contractual obligations due to the effects of the coronavirus outbreak (https://www.reuters.com/ article/us-china-health-trade/china-trade-agency-to-offer-firms-force-majeure-cer tificates-amid-coronavirus-outbreak-idUSKBN1ZU075) (Nwedu 2021). As such, if a business in China located in a zone with government-imposed quarantine has a force majeure clause in a contract governed by Chinese law, the invocation of a force majeure clause may be successful.

Other options: Impossibility/impracticability and purpose frustration

If a party cannot successfully use a force majeure clause to excuse performance during a coronavirus outbreak or if a contract does not contain a force majeure clause, other options may be available to excuse reduced performance, such as the defense of impossibility and impracticability (https://www.reuters.com/article/us-china-health-trade/china-trade-agency-to-offer-firms-force-majeure-certificates-amid-coronavirus-outbreak-idUSKBN1ZU075). The Uniform Commercial Code (U.C.C.) States that a seller is excused from performing under a contract when "performance under the agreement was rendered impracticable by the occurrence of an unforeseen event whose default was a basic assumption based on which the contract was concluded or by complying with good faith in any applicable external or internal government regulation or order whether subsequently proved invalid" (SHB.com 2020). The contract retreating defines the impossibility as "not only a strict impossibility but impracticability due to unreasonable difficulties, costs, injuries or losses involved" [Second Restatement of Contacts].

Suppose a contract does not contain a force majeure clause and a defense of the impossibility or impracticability fails. In that case, another possible defense for a party that cannot fulfill its obligations because of the coronavirus is the frustration of the purpose. In order for the doctrine to apply, "the frustrated aim must be so completely the basis of the contract that, as both parties understood, the transaction would have made little sense without it" (Crown I.T. Servs. v. Olsen 2004). In other words, purpose frustration occurs when "a change of circumstances makes the performance of one party virtually worthless for the other, frustrating its purpose in concluding the contract" (P.P.F. Safeguard, L.L.C. v. B.C.R. Safeguard Holding, L.L.C. 2011). Businesses should, however, be aware of economic difficulties because an increase in the cost of performance under a contract is not sufficient to support the frustration of defending the purpose (A + E Television Networks, L.L.C. v. Wish Factory Inc. 2016).

Coronavirus has a significant and damaging impact on companies and their ability to perform under the contracts they have agreed to. However, if a claimant party can successfully invoke a force majeure clause, a defense of impossibility/impracticability or frustration of the purpose defense to excuse performance because of coronavirus is an intense investigation in fact and should be assessed on a case-by-case basis. The parties to the contract should look at the contract's specific language, including the applicable law, to determine their likelihood of success (https://www.akerman.com/en/perspectives/the-coronavirus-and-force-majeure-clauses-in-contracts.html).

6.2 Force Majeure Versus Pacta Sunt Servanda

In general, force majeure conflicts with the concept of "pacta sunt servanda" (agreements must be preserved), a key concept in civil and international law with analogs in the common law. It is assumed that it is not easy to escape contractual liability, and it is difficult to prove that the events have been unforeseeable.

As science evolves, the world becomes aware of natural threats we previously did not know, such as solar explosions, asteroids, pandemics, super volcanoes, or earthquakes. We are also developing new human threats, such as cybernetic, nuclear and biological warfare capabilities. They raised questions about what is and is not "predictable" in the legal sense.

We are also becoming increasingly aware of human capacity to cause events that have generally been considered Acts of God, such as climatic and seismic events. Ongoing disputes explore whether drilling and construction projects that have contributed to natural disasters that have rendered them inoperable qualify as force majeure. In short, the concepts underlying the force majeure are changing.

6.3 Force Majeure Under Electricity Laws

Force majeure

Force majeure simply refers to an event beyond the reasonable control of a person or part of a contract as a result of which that party or person is unable to fulfill his contractual obligations. Black's Law Dictionary defined it as "an event or effect that cannot be anticipated or controlled". A contractual provision allocates the risk of loss if performance becomes impracticable or impracticable, mainly due to an event that the parties could not have anticipated or controlled (https://economictimes. indiatimes.com/small-biz/legal/what-is-force-majeure-the-legal-term-everyone-sho uld-know-during-covid-19-crisis/articleshow/75152196.cms).

Force majeure includes extreme natural phenomena, terrorism, war, earthquake, hurricane, acts of government, explosions, epidemics and a non-exhaustive list in which the parties to the contract cover other events considered force majeure.

Force majeure, as the name suggests, is a type of force or consequence over which man has no control or of events that cannot be anticipated. The term is widely used in contracts because it refers to any unforeseen event that cannot be held in advance, but if that event occurs, it is covered by that term. In order to apply the force majeure clause to an event, such an event must take place outside the reasonable control of the parts (https://www.toppr.com/guides/business-laws/indian-contract-act-1872-part-ii/contingent-contracts/).

6.3.1 The Difference Between the Doctrine of Frustration and Force Majeure

The doctrine of frustration, as defined in Section 56 of the Indian contract Act, 1872, shall mean any oversight event that has taken place after the performance of a contract or during the performance of the contract, over which the parties have no control, and the fulfillment of such a contract becomes impossible or illegal, then such contracts become null and frustrated (https://us.eversheds-sutherland.com/por talresource/USForceMajeureGuideUS.pdf).

The term contractual frustration was recognized for the first time by English law in the Taylor versus Caldwell judgment (https://en.wikipedia.org/wiki/The_Electrici ty_Act), in which the court considered that in contracts where performance depends on the continued existence of a given person or thing, performance resulting from losing the person or work is assumed to be impossible will excuse performance.

The reference to force majeure can also be found in section 32 of the Indian contract Act of 1872, which defines the contingent contract (https://www.indialega llive.com/special-story/post-covid19-power-gencoms-discoms-duel-force-majeure-94779). According to these types of contracts, performance by the promise is based on conditions that are practically the case or that do not occur in an event when such conditions are met; the promise meets its obligations.

The Commanders' order on the late increase in payments to gencoms was reduced by the Central Electric Energy Regulatory Commission (C.E.R.C.).

6.4   The Drafting of an Effective Force Majeure Clause

A contractual term that only States that the "normal force majeure clauses" apply was considered void for uncertainty, and therefore the parties must carefully consider how force majeure is intended to be applied to their contract. This is particularly important in the case of force majeure clauses, in which they are to be interpreted by reference to the express words used and not by the general intention of the parties (https://us. eversheds-sutherland.com/portalresource/USForceMajeureGuideUS.pdf).

According to the example below, a force majeure clause in English law will usually seek to exclude liability or to apologize for non-performance in certain circumstances (with a reasonable degree of specificity), followed by a general sentence: "No party shall be liable, except for compensation provided herein, for payment of sums due under this Agreement for failure to comply with the terms of the Agreement when enforcement is prevented or prevented by strikes (except for strikes imposed by the contractor's staff) or by blocking, revolting, armed conflict (declared or undeclared), insurrection, civil unrest, fires, interference by any governmental authority or other cause that such a party cannot reasonably control".

The essence of a "force majeure event" is predicting the unpredictable. This is a highly personal objective—a product of the parties' negotiations and lawyers' success in understanding their clients' business and associated risks. While noting the latter, some important drafting principles should be retained:

- First of all, synchronization. It is obviously unsatisfactory (and non-commercial) for the contractual obligations to remain suspended indefinitely. A valid force majeure event may terminate the contract as a matter of common law (Treitel 2014b). More frequently, however, force majeure clauses are drafted in such a way as to have a limited effect in time, which does not exceed total discharge. In order to ensure adequate protection for the specific requirements of each party, those force majeure clauses may wish to: and if terminated, provided that the party that terminated shall have a form of compensation (e.g., where advance payment has been made for services or goods not supplied).

- Secondly—to make a force majeure clause more valuable than the otherwise occurring frustrations and impracticability—force majeure events should be precisely defined to capture the risks specific to the industry. If negotiated, the economic difficulties can also be an event, but otherwise, it is unlikely to be considered part of a general catch-all provision.
- Thirdly, a force majeure clause functions as an exclusion clause, excusing a party from fulfilling its contractual obligations. Therefore, it can be submitted to the reasonableness test under Section 3 of the unfair contract terms Act of 1977 or the fairness and transparency requirements of the Consumer Rights Act of 2015. A force majeure clause drawn up too widely may be considered unreasonable and declared null and void. As such, it does not provide adequate protection to a party and leaves it subject to a claim.
- Fourthly, the parties should consider how the usual contractual relationship principles apply. That is, "the expression of one thing excludes other things", and the words listed are "of the same kind".

6.5 Force Majeure in Practice

6.5.1 What "Circumstances" Could a Force Majeure Clause Activate?

Force majeure events are commonly referred to as "Acts of God". However, the term's meaning has been expanded to cover much more than natural disasters such as wars and pandemics.

In Lebeagupin v. Richard Crispin, McCardie J. has carried out a helpful analysis of actions that can be considered "Acts of God" such as war, strikes, state actions (such as embargoes and denial of license); all of those can be considered force majeure (Lebeaupin v. Richard Crispin and Company 1920). The fact that certain events are not adequately characterized as force majeure events indicates the critical importance of a proper definition for relevant events to be mentioned in the written contract. According to English law, an event that makes the contract more expensive to perform will not usually be an "Act of God" and it will not lead to the conclusion of the "impossible to fulfill" (https://www.stiveschambers.co.uk/content/uploads/2020/08/Revisiting-Force-Majeure-by-Karamjit-Singh.pdf).

It is also essential that the alleged force majeure event is the sole cause of a party's alleged failure to fulfill its contractual obligations. In Seadrill Ghana operations Ltd against Tullow Ghana Ltd (2018), the defendant, Tullow, sought to invoke a force majeure clause because of a "drilling moratorium" imposed by the Government of Ghana. Seadrill claimed that Tullow is terminated for convenience reasons, as a collapse in oil prices made Seadrill's hiring drilling rigs less commercially attractive.

Finally, the court held that—although the drilling moratorium was an event of force majeure established on the facts—this was not the only cause of Tullow's failure to fulfill its contractual obligations.

The alleged termination for convenience is also well illustrated by Thames Valley Power Ltd versus Total gas & Power Ltd, if a force majeure clause in a gas supply contract—which provides for the release of contractual obligations in the event of failure to perform—was not considered to affect performance and had only become "economically more burdensome" (Thames Valley Power Ltd versus Total Gas & Power Ltd 2005). Consequently, the parties cannot generally hope to invoke force majeure to escape from the burden of a contract that remains physically and legally possible, albeit unprofitable or less profitable.

6.5.2 *"Reasonable Control", Causality and Mitigation*

The English law does not stipulate that the relevant circumstances invoking force majeure are unforeseeable; instead, the burden is on the invoking party to demonstrate circumstances beyond its reasonable control (or that of its agent). Crudesky found that "reasonable control" ultimately means exercising "one choice rather than another". In this respect, the problem of "fault" is neither decisive nor relevant. A crucial question in assessing a party's "reasonable control" is how closely it is causally linked to the force majeure event to the non-execution of the invoking parts. This is specific to the fact and depends on the language of the clause (https://us.eversheds-sutherland.com/portalresource/USForceMajeureGuideUS.pdf).

As a matter of English law, what is clear is that the alleged force majeure event must be the only cause of a party's failure to execute.

Where relevant, a Court will also often look to alternative performance modes as it did recently in Classic Maritime v. Limbungan Makmur (2018). There, the possibility of transport from an alternative port was an important factor in considering whether an explosion dam constituted an event of force majeure. Therefore, if alternative ways of performance are available and are not adequately explored, it is less likely that the court would consider the non-execution of a party attributable only to the alleged force majeure event.

6.6 Invoking Force Majeure in a Crisis

In 2011, Fukushima-Daiichi, Japan, faced a disaster at the nuclear power plant, leaving Tokyo Electric Power Company (TEPCO) with no option but to invoke force majeure and conclude its uranium supply agreement with the Canadian uranium mining company, Cameco. However, TEPCO also rejected the claim and had to pay USD 40 million in damages to the supplier Cameco (https://www.gatewayhouse.in/force-majeure-crisis/).

However, its interpretation changes in line with the situation and the legality. Epidemics such as SARS (2003), H1N1 (2009), and Ebola (2013) have seen several companies exempt from invoking force majeure clauses in commercial contracts. The French judiciary, for example, has decided to exclude epidemics such as H1N1 as force majeure events. However, in March 2020, one of the first COVID-19 judgments included COVID-19 as a force majeure event when a detainee could not attend a trial due to exposure to the virus (Baker McKenzie 2020).

In this spirit of analysis, the hydrological drought in a hydrographic basin that supplies water to a hydropower plant can be invoked as a cause of force majeure, if the estimated electricity production can no longer be achieved.

However, if the phenomenon is repeated, with a certain reverence, can it still be regarded as an unpredictable event?!

How can companies seek exemption from contracts and mitigate their losses during such events or pandemics?

The first thing to know is that each country has a different way of applying this measure. In countries like China, it is governed by the country's statute. In the current pandemic, the China Council for the Promotion of International Trade has issued more than 5000 force majeure certificates to help suppliers with foreign trading partners (Arnold & Porter 2020). However, the validity of these certificates for enforcement against a foreign partner is questionable and depends on the law governing a contract. A legislative clause sets out the laws of the country that will apply to a contract.

In other countries, such as India, the United Kingdom and Japan, force majeure clauses are part of individual private contracts and differ according to how each contract was classified. Even in the same industry or supply chain, two companies from the same companies may have different force majeure clauses. This is particularly worrying for buyers and suppliers, as applying force majeure by a link affects the whole supply chain.

How are the competencies of the authorities issuing certificates that establish force majeure determined at state level?!

At present, for example, India does not have a clear, codified law on force majeure (Ronen 2010). The Indian Contract Act, 1872 contains two provisions relevant in the case of a dispute on force majeure or the absence of a force majeure clause in a contract. Section 32 addresses enforcement of contingent contracts, and Section 56 provides a "doctrine of frustration" which provides that if the objective of the contract has become impossible to perform or an event makes the performance impossible to perform, then the contract can be deemed void (https://www.gatewayhouse.in/force-majeure-crisis/). The Ministry of Finance's Manual for the purchase of goods 2017 (Ministry of Finance, Government of India 2017) contains a force majeure clause for suspending performance under a contract but does not explicitly cover government actions or epidemics. However, in light of COVID-19, the Indian government observed and changed this quickly. In February 2020, the Ministry of Finance stated that any disruption of supply chains due to the unavailability of components imported from China or other countries would be treated as a force majeure event (Ministry of Finance,

Government of India 2020). This provides immediate relief to government contracts with international suppliers. India imports more than 80% of its solar modules and cells from China (Verma 2020). The Ministry of New and Renewable Energy notified the industry companies to consider the pandemic and the supply chain interruptions a force majeure event (Ministry of New and Renewable Energy 2020).

At the moment, many Indian companies are scanning their contracts to find force majeure clauses, looking for ways to mitigate losses. From a legal point of view, it can have three possible outcomes:

Invoke force majeure: In March 2020, Hero MotoCorp invoked force majeure clauses in its contracts with suppliers and suspended payment (Economic Times Bureau 2020a). Similarly, Reliance Retail: Future Group, P.V.R. sent notifications to invoke force majeure clauses in their lease contracts (Economic Times Bureau 2020b). PetroChina, China's largest gas producer and pipeline gas supplier, has issued force majeure notifications, thereby suspending natural gas imports (Aizhu and Jaganathan 2020). The largest importer of liquefied natural gas (LNG) in China, China National Offshore Oil Corporation, also submitted force majeure notices to suppliers such as Total and Shell, which were rejected (S&P Global Patts 2020).

1. Renegotiation of contracts: This can be done when the force majeure clause does not provide for an adequate remedy. However, even when force majeure is appropriate, companies influenced by events such as droughts, landslides or the COVID-19 pandemic may renegotiate contracts. Renegotiation, however, requires mutual consent.
2. Dispute resolution: The only legal recourse available is adopting a dispute settlement mechanism when a contract does not contain a force majeure clause or where the clause is in dispute. It may be by mediation or arbitration.

References

A + E Television Networks, L.L.C. v. Wish Factory Inc. (2016) WL 8136110, at *13 (S.D.N.Y. Mar. 11, 2016)

Aizhu C and Jaganathan J (2020) PetroChina suspends some gas contracts as coronavirus hits demand. Reuters, 5 Mar 2020. https://in.reuters.com/article/petrochina-gas/exclusive-petrochina-suspends-some-gas-contracts-as-coronavirus-hits-demand-sources-idINKBN20S10J

Arnold & Porter (2020) What to do when you receive a coronavirus-related Force Majeure notice, 4 Mar 2020. https://www.arnoldporter.com/en/perspectives/publications/2020/03/what-to-do-when-you-receive-a-coronavirus

Article 1218 of the French Civil Code re-defined force majeure on Oct. 1 2016. For the first time, it codified the three essential elements that comprise force majeure: (i) externality (l'extériorité); (ii) unforeseeability (l'imprévisbilité); and (iii) inevitability (l'inévitabilité). The French concept of force majeure, inherited from a Roman law desire to restrict strict liability, centered initially around the quite restrictive concept of 'irresistibility' but under Article 1218 has moved to a more flexible concept of 'inevitability' (e.g., whether the effects were unavoidable) to respond to the demands of international trade

Baker McKenzie (2020) France: first decision to declare COVID-19 outbreak as a Force Majeure event, 31 Mar 2020. https://www.bakermckenzie.com/en/insight/publications/2020/03/france-decision-declare-covid19-force-majeure

Bloom v. Home Devco/Tivoli Isles, L.L.C. (2009) WL 36594, at *4 (S.D. Fla. Jan. 6, 2009) (quoting Florida Power Corp. v. City of Tallahassee, 18 So.2d 671, 675 (Fla. 1944)); see also Fru-Con Const. Corp. v. U.S., 44 Fed. Cl. 298, 314 (1999)

Classic Maritime v. Limbungan Makmur (2018) EWHC 2389, at [66]. This is due to be heard on appeal on 11 June 2019

Commonwealth Edison Co. v. Allied-Gen. Nuclear Servs. (1990) 731 F. Supp. 850, 859 (N.D. Ill. 1990)

Coronavirus COVID 19: facts and insights (Updated: March 9, 2020)—global health + crisis response, by McKinsey & Company

Crown I.T. Servs. v. Olsen, 11 A.D.3d 263, 265 (1st Dep't. 2004).

Economic Times Bureau (2020a) Hero MotoCorp suspends payments to suppliers amid lockdown, 30 Mar 2020a. https://economictimes.indiatimes.com/industry/auto/two-wheelers-three-whe elers/hero-motocorp-suspends-payments-to-suppliers-amid-lockdown/articleshow/74877047. cms?from=mdr

Economic Times Bureau (2020b) Lockdown effect: restaurants, cinemas & retailers at malls seek zero rentals till May, 3 Apr 2020b. https://economictimes.indiatimes.com/industry/services/ retail/lockdown-effect-restaurants-cinemas-retailers-at-malls-seek-zero-rentals-till-may/articl eshow/74956239.cms?from=mdr

El Paso Field Servs., L.P. v. Mastec N. Am., Inc. (2012) 389 S.W.3d 802, 808 (Tex. 2012)

General Contract Clauses: Force Majeure, Practical Law Standard Clauses 3-518-4224; see also Kel Kim Corp. v. Cent. Markets, Inc., 70 N.Y.2d 902 (1987); Rochester Gas & Elec. Corp. v. Delta Star, Inc., 2009 WL 368508, at *2 (W.D.N.Y. Feb. 13, 2009)

Hennings WC, Abdellatif SA, Hanna AS (2022) Proper risk allocation: force majeure clause. J Legal Affairs Dispute Resol Eng Constr 14(1), Article Number 04521048. https://doi.org/10. 1061/(ASCE)LA.1943-4170.0000527

https://economictimes.indiatimes.com/small-biz/legal/what-is-force-majeure-the-legal-term-eve ryone-should-know-during-covid-19-crisis/articleshow/75152196.cms (2017) 14 SCC 80–22 ER 309; 3 B. & S. 826 (1863)

https://en.wikipedia.org/wiki/The_Electricity_Act

https://us.eversheds-sutherland.com/portalresource/USForceMajeureGuideUS.pdf

https://www.akerman.com/en/perspectives/the-coronavirus-and-force-majeure-clauses-in-contra cts.html

https://www.cdc.gov/flu/pandemic-resources/1918-pandemic-h1n1.html

https://www.cdc.gov/flu/pandemic-resources/1957-1958-pandemic.html

https://www.cdc.gov/flu/pandemic-resources/2009-h1n1-pandemic.html

https://www.gatewayhouse.in/force-majeure-crisis/

https://www.indialegallive.com/special-story/post-covid19-power-gencoms-discoms-duel-force-majeure-94779

https://www.jdsupra.com/legalnews/the-coronavirus-and-force-majeure-56194/

https://www.law.cornell.edu/wex/force_majeure

https://www.lawgazette.co.uk/legal-updates/force-majeure-in-2020/5103626.article

https://www.linkedin.com/pulse/force-majeure-deepak-dayal-1c/

https://www.reuters.com/article/us-china-health-trade/china-trade-agency-to-offer-firms-force-majeure-certificates-amid-coronavirus-outbreak-idUSKBN1ZU075

https://www.stiveschambers.co.uk/content/uploads/2020/08/Revisiting-Force-Majeure-by-Kar amjit-Singh.pdf

https://www.toppr.com/guides/business-laws/indian-contract-act-1872-part-ii/contingent-contra cts/

Jester BJ, Uyeki TM, Jernigan DB (2020) Am J Public Health 110(5):669–676. Published online 2020 May. https://doi.org/10.2105/AJPH.2019.305557

Lebeaupin v. Richard Crispin and Company (1920) 2 K.B. 714, 702

Levashova Y (2020) Fair and equitable treatment and investor's due diligence under international investment law, Netherlands. Int Law Rev 67(2):233–255. https://doi.org/10.1007/s40802-020-00170-7

Milovanovic P, Dumic I (2021) COVID-19 as a "Force Majeure" for non-COVID-19 clinical and translational research. Comment on "Analysis of scientific publications during the early phase of the COVID-19 pandemic: topic modeling study. J Med Internet Res 23(5), Article Number 27937. https://doi.org/10.2196/27937

Ministry of Finance, Government of India (2017) Manual for procurement of goods 2017. https://doe.gov.in/sites/default/files/Manual%20for%20Procurement%20of%20Goods%202017_0_0.pdf

Ministry of Finance, Government of India (2020) Office memorandum—Force Majeure, 19 Feb 2020. https://doe.gov.in/sites/default/files/Force%20Majeure%20Clause%20-FMC.pdf

Ministry of New and Renewable Energy (2020) Office memorandum: time extension in scheduled commissioning date of RE projects considering disruption of the supply chains due to spread of coronavirus in China or any other country as Force Majeure, 20 Mar 2020. https://mnre.gov.in/img/documents/uploads/file_f-1584701308078.pdf

Nwedu CN (2021) The rise of Force Majeure amid Coronavirus pandemic: legitimacy and implications for energy laws and contracts. Nat Resour J 16:1

O'Reilly C (2021) Violent conflict and institutional change (1). Econ Trans Inst Change 29(2):257–317. https://doi.org/10.1111/ecot.12269

P.P.F. Safeguard, L.L.C. v. B.C.R. Safeguard Holding, L.L.C. (2011) 924 N.Y.S.2d 391, 394 (quoting Restatement (Second) of Contracts § 265, Comment a)

Perlman v. Pioneer Ltd. P'ship, 918 F.2d 1244, 1248 (5th Cir. 1990) ("Because the clause labeled "force majeure" in the lease does not mandate that the force majeure event be unforeseeable or beyond the control of [the non-performing party] before performance is excused, the district court erred when it supplied those terms as a rule of law."); See Sun Operating Ltd. P'ship v. Holt, 984 S.W.2d at 288 ("Indeed, to imply an unforeseeability requirement into a force majeure clause would be unreasonable.")

Phibro Energy, Inc. v. Empresa De Polimeros De Sines Sarl (1989) 720 F. Supp. 312, 318 (S.D.N.Y. 1989) (citing Kel Kim Corp. v. Central Markets, Inc., 70 N.Y.2d at 902-03)

Rochester Gas & Elec. Corp. v. Delta Star, Inc. (2009) WL at *7; Phibro Energy, Inc. v. Empresa De Polimeros De Sines Sarl, 720 F. Supp. at 318; see also Goldstein v. Orensanz Events LLC, 146 A.D.3d 492, 493 (1st Dep't 2017)

Ronen Y (2010) Incitement to terrorist acts and international law. Leiden J Int Law 23(3):645–674. https://doi.org/10.1017/S0922156510000269

Seadrill Ghana Operations Ltd v. Tullow Ghana Ltd (2018) EWHC 1640

Second Restatement of Contacts § 254

SHB.com (2020) Force Majeure and common law defenses. A national survey

S&P Global Patts (2020) China's CNOOC declares force majeure on LNG contracts amid coronavirus outbreak, 6 Feb 2020

Sun Operating Ltd. P'ship v. Holt (1998) 984 S.W.2d at 283–284

Thames Valley Power Ltd v. Total Gas & Power Ltd (2005) EWHC 2208

Treitel (2014a) Impractibility in English law—exceptional or special cases—express provisions. In: Frustration and Force Majeure, 3rd ed. §6-036

Treitel G (2014b) 12-020—Provisions for non-frustrating events, purpose and nature. In: Frustration and Force Majeure, 3rd ed

Verma A (2020) India imported solar equipment worth $1.18 Bn from China between April–December, 6 Mar 2020. https://www.saurenergy.com/solar-energy-news/india-imported-solar-equipment-worth-1-18-bn-from-china-between-april-december

Chapter 7
Sample Contractual Provisions in a Force Majeure Contract

Abstract Force majeure is a phrase used in commercial contracts that highlights the impossibility of providing the services assumed by the contract due to unforeseen causes, independent of the will of the signatory parties. This chapter presents the significance of this concept, of its way of writing and invoking in critical situations. The proper drafting of such a contractual provision can make the difference between concluding a contract in borderline situations without involving the parties in a dispute in court or arbitration.

Keywords Contractual provision · Model contractual · Force majeure

7.1 Introduction

The extreme events in the past period, but in particular the economic implications, have led to major global economic losses for the governments and companies (Ekmekci et al. 2020; Zin et al. 2021; Osipova and Eriksson 2011). In these circumstances, the management of the companies has realized how important the provisions of a contract are regarding the application of force majeure (Hennings et al. 2022; Casady and Baxter 2020; Nocera 2020).

Force majeure contractual clauses generally provide for a postponement of the deadline for the performance of contractual tasks. In the absence of a force majeure clause, other clauses may or may not allow for recovery of time and money (e.g., no-damages clauses for delays in the supply of subassemblies or armed conflicts) (Nwedu 2021). In the absence of a force majeure clause in the contract, the common law principles of impossibility, impracticability and frustration may apply. This can provide a party with a measure of protection, including relief from the performance of all or part of the contractual obligations.

There is no standard force majeure clause to be used in contracts so as to fit all cases. However, the effort put into drafting a force majeure clause should be proportionate to the level of risk in the contact. Once a variant has been found in the drafting of such a clause, it will be continuously refined and adapted to new borderline situations that may arise in the course of the contract.

© The Author(s), under exclusive license to Springer Nature Switzerland AG 2023
D. C. Diaconu, *Force Majeure in the Hydropower Industry*,
https://doi.org/10.1007/978-3-031-27402-2_7

The clauses drafted will be adapted with regard to the concept and definition of force majeure in the specific region where they apply, the characteristics of events that may be considered as force majeure and the obligations of the parties to the contract when such events occur (Ezeldin and Abu Helw 2018; Hooshyar et al. 2019; Gazmuri and Olivares 2020; Hennings et al. 2022).

In drafting the clauses contained in this essential article, the following aspects should be taken into account to ensure the completeness and comprehensiveness of the force majeure clause:

- defining force majeure;
- defining events to be considered force majeure;
- establishing a way of notifying force majeure;
- establishing the obligation to mitigate the effects of the force majeure;
- defining the consequences of force majeure.

In their proposed and assumed contracts, many law firms or companies provide an article showing force majeure and the conditions of its invocation. Such an article may, without pretensions to be complete or most comprehensive, be presented as follows.

7.2   Content of the Force Majeure Clause

The content of the force majeure clause must take into account several milestones leading to the achievement of the final goal. In other words, drafting an article that provides a fair solution to an unforeseeable situation for all parties to the contract. In many situations, the process of negotiating contract terms between the parties leads to a fair contract for both parties. In the hydropower sector, as in other specific areas of activity, the content must be adapted to the specific problem.

In this drafting/negotiation process, a few benchmarks should be kept in mind.

7.2.1   Availability of Land and Access

- The investor generally takes the risk of choosing the site and acquiring the necessary land. Studies (geotechnical, hydrological, economic, etc.) are carried out to minimize the risk that the investment will not be completed or will deteriorate rapidly. However, unforeseen situations may arise, such as land retrocession or changes in land use that result in a ban on construction (natural sites, buffer areas, etc.). There are situations where investments are carried out in areas that are difficult to access, which may restrict access to machinery or equipment needed to maintain/upgrade production facilities, so that access to the site is the one that can generate force majeure situations.

- Site risks may also arise from the accumulation of impacts within a catchment where there are several hydropower developments, which for example may affect biodiversity. In these situations, additional land should be provided to ensure compensatory measures in this respect.

7.2.2 Design Risk

- Nowadays, scientific development, the use of computer technology and artificial intelligence make the design process of a hydropower construction process safer. Obviously, it is not possible to achieve 100% safety for such a project. At this stage of investment development, problems may arise in the technical and economic sustainability of the project, the approval of financing and construction, and in making changes to the project.

7.2.3 Environmental Risk

Environmental hazards have become much more common as public awareness of the need to protect the environment has increased. Many environmental non-governmental organisations are active at global or local level and can stop the implementation and completion of an investment at certain stages of its development. In many major projects, environmental permits are an essential component of the project and can take a long time to be prepared and approved both in the preconstruction, construction and commissioning phases. Risk categories can be:

- Preexisting conditions (proximity to protected areas, protected species in the area, etc.).
- Obtaining environmental permits (complexity of environmental legislation and pace of permitting).
- Compliance with Environmental Permits and Laws (Parties should ensure that the change in law provisions adequately address changes in (mandatory) environmental standards and laws to avoid disputes as to which party bears the consequences of any requirements imposed after the contract is signed. See also risk of change in legislation).
- Environmental conditions caused by the project. The implementation of large hydropower projects involves significant material and energy consumption. In these situations, measures are taken to recover the waste generated and recycle it in order to reduce the negative environmental impact. The construction of the project itself can generate environmental damage over different periods of time.
- External environmental events. It is not only the construction of an objective that can affect the environment, but also extreme natural phenomena that can lead to obstructions to the implementation works. Situations such as heavy rainfall

leading to high water flows or landslides, extreme air temperatures preventing concrete pouring, and many others have occurred over time.

- Climate change (the current climate evolution has shown that more attention needs to be paid to climate change events, taking into account the risk and impact of climate risk events on water resources as well as infrastructure (change in rainfall regime, increase in air temperature, occurrence of landslides, decrease in water flows, etc.).
- It may be appropriate to treat certain events as force majeure events if they occur above certain thresholds (e.g., temperatures exceeding certain ranges, water flows falling below certain values). Design resilience is also an important mitigation factor, for example, for projects with seasonal weather such as monsoon, or where earthquakes are frequent in the area. But by adapting the project to such factors with an extreme manifestation can generate an oversizing of the investment making the return on investment decrease or even become unprofitable.

7.2.4 Construction Risk

This stage involves a wider range of risks that may arise at certain points in the construction process. The complexity of the materials, machinery and technologies involved in the construction process can lead to some syncopations in the chain of production, transport, commissioning and operation.

- Rising costs, a situation generated either by economic considerations (inflationary moments) or by the lack of necessary material resources whose prices become prohibitive. There are also situations where project modifications are necessary and additional funding is required. However, these are included in the general estimate under the category of miscellaneous and unforeseen costs, which may amount to 10 or even 20% of the value of the investment.
- Delays in the completion of works, and especially in the delivery of equipment that is supplied by specific manufacturers (energy turbines, high-power transformers, concrete or steel prefabricated units, etc.).
- Project management and interface with other works/facilities.
- Quality assurance and other regulatory standards in the field of construction (as a rule, the builder is the first to carry out the tests necessary for quality certification through its own laboratories for the analysis of the quality of materials and their implementation, but the state, through its representatives in this field of construction quality, also carries out inspections at certain stages of project implementation).
- Compliance with occupational health and safety rules.
- Liability for death, personal injury, property damage and third-party liability
- Intellectual property—The contractor assumes the risk of obtaining all relevant licenses for the construction and operation of the hydropower project, unless the Contracting Authority imposes certain design or other technological solutions on the Private Partner, in which case the corresponding risk may be shared or borne by

the Contracting Authority. In many cases, due to the particularity of the project, the terrain where it is to be carried out, it has been necessary to identify new technical procedures, construction technologies, IT systems to meet the requirements. The regulation of their intellectual property needed to be clarified from the outset.

- Vandalism. Vandalism may pose a greater risk if the political climate opposes the hydropower project. Or in situations where there are project implementation stoppages, with long periods when construction is under conservation.

7.2.5 Social Risk

- Community and business. The contracting authority may choose to adopt internationally recognized social and environmental standards and practices for the project to manage social risk, particularly if international funding options are desirable.
- Relocation of communities. Hydropower constructions, involving the construction of large-scale developments or the formation of reservoirs, often require the relocation of communities (housing, cultural, educational and health facilities), transport routes or other culturally, religiously or naturally valuable sites. Depending on the ownership of land and buildings, unforeseen situations may arise in the expropriation process.
- Heritage/indigenous people. Represent in some countries risks that are difficult to anticipate or manage especially by foreign investors
- Industrial stocks. Risk of labour disputes and strikes adversely affecting the investment's progress, unless these actions fall under political risks—the contracting authority may bear the risk or share the risk (as a force majeure or relief event) for strikes and other large-scale labour dispute events.

7.2.6 Operational Risk of the Hydropower Objective

- Increased operating costs and impaired performance. Increased costs and delays in the operational phase can have a variety of causes, from errors in maintenance cost estimates to extreme weather events.
- Performance risk/price risk.
- Risk related to operational resources or inputs. Hydrological risk is the main input risk for hydropower projects once they become operational. Variation in water volumes and water quality (especially sediment that clogs water storage reservoirs) should be assessed by reference to detailed and accurate historical records over as long a period as possible. Some of these water quality and sediment risks can be controlled by appropriate watershed management (maintenance of forested areas, slope management, etc.).

- Intellectual property. The implementation of new exploitation, forecasting or other programs will be carefully regulated in this respect.
- Health and safety compliance.
- Maintenance standards.

7.2.7 Risk of Material Adverse Government Action (MAGA)

In projects where a MAGA provision is appropriate, the contracting authority assumes the risk that certain "political" actions may have a significant negative effect on the private partner's ability to fulfil its contractual obligations or on its rights or financial status (https://www.lawinsider.com/dictionary/material-adverse-government-action). The contracting authority is responsible for costs and delays and is usually exposed to the risk of termination in the event of prolonged MAGA events. Although not all jurisdictions use the term "MAGA", many have equivalent provisions under different terminology. This type of risk is all the greater as the investing company is financially and administratively controlled by a state entity, which may be directly influenced by political decisions (appointment of manager, budget allocation, taxation of profits, etc.). Changing legislation affecting the construction or operation of the hydropower project can also be considered a MAGA action.

7.3   Invocation of the Force Majeure Clause

When invoking the force majeure clause, there is a path to follow with several milestones to meet. Upon the occurrence of a force majeure event, the affected party shall promptly notify the other party of the occurrence of the force majeure event, its effect on the performance of the contract and its expected duration. The evolution of the triggering event must be closely monitored and regular information of the parties involved must be carried out. Failure by the affected party to give such notification in reasonable time shall entitle the other party to compensation for losses suffered. If the triggering event of the extreme situation ceases, then the contract can be resumed and the contract can be terminated.

If the critical situation is prolonged in duration and intensity, the inability to perform the contract either temporarily or permanently will result. In the case of a permanent situation, the affected party who has become unable to perform the contractual provisions must notify the other party and request the termination of the contract with clarification of the financial situation between the parties. Temporary suspension of the contractual terms subsequently allows the resumption of activities and the conclusion of the contract (Fig. 7.1).

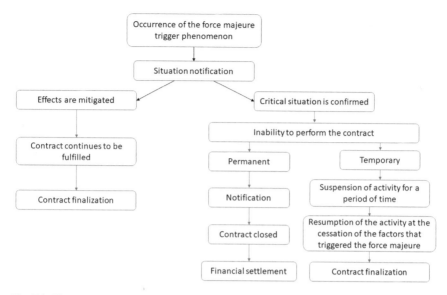

Fig. 7.1 Flowchart of the occurrence and consequences of force majeure situations

7.4 Conclusions

The development of a sample force majeure clause can be done by a review of the literature, including identifying force majeure as a legal concept and its approach in both the civil and common law systems, but also by consulting firms on force majeure as a concept and the provisions that are recommended to be included in contracts as well as law firms dealing with commercial contracts.

It is concluded that force majeure is a civil law concept aimed at excusing parties to a contract from their obligations upon the occurrence of events which are beyond the control of the parties and are neither foreseeable nor avoidable. Such events can be natural disasters, acts of state or governmental action, military or civil military action, acts of terrorism or war, and nuclear disasters.

References

Casady CB, Baxter D (2020) Pandemics, public-private partnerships (PPPs), and force majeure|COVID-19 expectations and implications. Constr Manag Econ. https://doi.org/10.1080/01446193.2020.1817516

Ekmekci E, Gunes G, Kasikci M, Gumuskaya G (2020) A critical approach and recommendations on force majeure and its legal consequences in tax law. Istanbul Hukuk Mecmuasi 78(2):1069–1138. https://doi.org/10.26650/mecmua.2020.78.2.0026

Ezeldin AS, Abu Helw A, (2018) Proposed force Majeure clause for construction contracts under civil and common laws. J Legal Aff Dispute Resolut Eng Constr 10(3).https://doi.org/10.1061/(ASCE)LA.1943-4170.0000255

Gazmuri ID, Olivares AV (2020) The effects of contractual non-performance caused by force majeure. Revista de Derecho Civil 7(3):123–161

Hennings WC, Abdellatif SA, Hanna AS (2022) Proper risk allocation: force majeure clause. J Legal Aff Dispute Resolut Eng Constr 14(1), Article Number 04521048. https://doi.org/10.1061/(ASCE)LA.1943-4170.0000527

Hooshyar MR, Zare M, Khosravinia B (2019) The impact of force majeure on non-contractual liability in Iranian law. Dilemas Contemporaneos-Educacion Politica Y Valores, 6, Special Issue SI, Article Number 68

https://www.lawinsider.com/dictionary/material-adverse-government-action. Accessed 20 Dec 2022

Nocera J (2020) Get ready for a tsunami of pandemic lawsuits over broken deals [online]. Bloomberg businessweek. Available from: https://www.bloomberg.com/news/articles/2020-05-04/pandemic-lawsuits-will-ask-is-the-coronavirusan-act-of-god. Accessed 14 May 2022

Nwedu CN (2021) The rise of force majeure amid the coronavirus pandemic: legitimacy and implications for energy laws and contracts. Nat Resour J 61(1):1–18

Osipova E, Eriksson PE (2011) How procurement options influence risk management in construction projects. Constr Manag Econ 29(11):1149–1158

Zin SM, Rahmat NE, Razak AMA, Fathi NH, Budiartha INP (2021) A proposed pandemic clause for force majeure events under construction contracts in Malaysia. Environ Behav Proc J 6(16):33–37. https://doi.org/10.21834/ebpj.v6i16.2733

Chapter 8
Concrete Cases of Invocation of Force Majeure

Abstract Invoking force majeure is not unusual, at least recently, so we briefly presented some large-scale hydroenergy situations. News agencies' specialized reports made public highlight such situations. In general, these situations are either avoided because the companies operating in this industry are financially strong, or it is precisely because of the high investment in the construction and facilities needed to produce electricity.

Keywords Extreme situations · Failure

8.1 Private Power Producer Disputes British Columbia Hydro's Cancellation of $20 Million in Purchases

Private power firm Innergex says British Columbia Hydro has involved "force majeure" on six of its British Columbia plants.

Canadian utility British Columbia Hydro operating as B.C. Hydro canceled $20 million in electricity purchases from six private power plants. It said the company that operates them and disputes the reasons for B.C. Hydro which refuses to supply electricity (https://vancouversun.com/business/private-power-producer-disputes-b-c-hydros-cancellation-of-20-million-in-purchases).

B.C. Hydro cites the current COVID-19 pandemic and the associated government measures taken in response to it as a force majeure event, which means it is an event that cannot be controlled by B.C. Hydro, which prevented the company from fulfilling its contract, Innergex said in a press release (https://vancouversun.com/business/private-power-producer-disputes-b-c-hydros-cancellation-of-20-million-in-purchases).

Innergex will comply, the company said, "but it will do so as a protest and will try to exercise its rights" under its contract with B.C. Hydro.

B.C. Hydro said on the 11 May that it would consider reducing energy purchases from independent power producers in order to cope with a sharp drop in electricity demand associated with the COVID-19 pandemic.

B.C. Hydro and Power Authority is a Canadian electricity company in the province of British Columbia, generally known only as B.C. Hydro. It is the main electricity

© The Author(s), under exclusive license to Springer Nature Switzerland AG 2023   85
D. C. Diaconu, *Force Majeure in the Hydropower Industry*,
https://doi.org/10.1007/978-3-031-27402-2_8

distributor, serving 1.8 million customers. As a provincial corporation of the Crown, B.C. Hydro reports the Ministry of Energy and Mining of British Columbia and is regulated by the British Columbia Utilities Commission (BCUC). B.C. Hydro operates 32 hydroelectric installations and three thermal power plants supplied with natural gas. Since 2014, 95% of the province's electricity has been produced by hydroelectric power plants, which mostly consist of large hydroelectric dams on the rivers Columbia and Peace. The different plants of B.C. Hydro generate between 43,000 and 54,000 GW-h electricity each year, depending on prevailing water levels. The electricity produced is supplied through a network of 18,286 km of transmission lines and 55,254 km of distribution lines. For fiscal year 2013–2014, domestic electricity sales reached 53,018 GWh, with revenues of $5392 billion and net revenue of $549 million (https://wikimili.com/en/BC_Hydro).

This coincided with the melting of spring snow, which usually fills the water tanks in British Columbia. The energy producer B.C. Hydro faces the situation of managing the fall in energy demand and the increase in water volumes in the hydrographic network.

B.C. Hydro's spokeswoman, Mora Scott, said they had to stop three of their own smaller power plants, namely Shuswap Falls, Fall River near Prince Rupert and Aberfeldie near Cranbrook.

"I think it would have been beneficial to consult regarding a way of solving the problems", said Clean Energy B.C. Managing Director Laureen Whyte before resorting to force majeure (BC Hydro 2014).

8.2   The Destruction of a Dam in Laos

In Laos, in 2018, 23 July, the Saddle D dam collapsed, part of a large hydroelectric system of dams under construction in the Champasak Province of the southeast of Laos. The collapse of the dam led to the widespread destruction and homelessness of the local population in the neighboring Attapeu Vientiane Province of 29 May (Xinhua). According to the local daily Vientiane Times, this deterioration of the dam built within the XE-Pian XE-Namnoy hydropower project in the southern province of Attapeu in Laos was not considered a "major force", which quoted a senior government official (https://www.rfa.org/english/news/laos/soil-052820191 53902.html).

Singphet Bounsavatthhanh, the vice president of the National Investigation Committee (CNI), which was formed by the Lao Government to investigate the cause of the disaster, which killed dozens of citizens and left thousands of other citizens homeless, revealed the results of the investigation in a press conference.

Singphet said that even if the rainfall was quite abundant in the days before the tragedy of 23 July 2018, the water level in the reservoir is still below the maximum working level and well below the crest level (canopy) when the collapse began. He said: The incident cannot be considered a "force majeure", he told local media.

The Lao Government has established the NIC to carefully investigate, verify and report to the Lao Government, the general public, the international community and other stakeholders on the underlying causes of the dam collapse.

In order to ensure transparency in the investigation of the causes of the collapse, the committee engaged and gave full authority to a globally recognized professional and independent expert group (IEP) to conduct an independent investigation into the accident and to report on the results of this investigation. In addition, the Lao Government invited foreign parties to participate as observers in the investigation process.

The Government of Laos understands that the Korean company involved in the development of the project is also conducting an investigation into the causes of the collapse. The IEP presented its conclusions on the failure of the dam in a final report and presented its report to the INC on 20 March 2019. It found that the main cause of the failure was linked to the high permeability of the foundation ground combined with eroded horizons.

As the water level increased during tank filling, the flow of infiltrations developed into the foundation along these paths and horizons with high permeability.

This led to internal erosion and softening of the soil side, which reached a certain critical point, the stability of the dam was no longer ensured, and a deep rotational slip was triggered in the highest section of the dam.

The dam's collapse led to devastating floods, leaving 42 dead and 23 missing, the Lao Government reported last December (Fig. 8.1) (http://www.esa.int/Applic ations/Observing_the_Earth/Copernicus/Sentinel-1/Sentinel-1_maps_flash_floods_ in_Laos).

The International Rivers Organization—which Washington Post described as "a non-governmental group that is generally critical for such projects"— suggested that the collapse illustrates the "major risks" involved if the construction is "unable to cope with extreme weather conditions" "because, especially in Laos, unpredictable and extreme weather events are becoming more and

Fig. 8.1 Aerial imagery after dam failure (http://www.esa.int/Applications/Observing_the_Earth/ Copernicus/Sentinel-1/Sentinel-1_maps_flash_floods_in_Laos)

more frequent". Indeed, there was an escape order for the downstream area of the dam because of the discovery of cracks in it. These damages had been reported to the company by South Korean entrepreneurs "at least one day before" the floods (https://www.internationalrivers.org/wp-content/uploads/sites/86/2020/08/ir-factsheet-2_year_xe_pian_dam_collapse_1_0.pdf, https://www.rfa.org/english/news/laos/soil-05282019153902.html).

8.3 Oroville Dam

The Oroville dam is an important part of the California State Water Project. It is an earth ground dam on the Feather River, east of the town of Oroville in Northern California. The water accumulation is used for flood control, water storage, electricity generation and water quality improvement in the Sacramento–San Joaquin River Delta. Built in 1968, it is the tallest dam in the United States, at 230 m. The lake formed is capable of storing more than 4.3 billion m^3. The adjoining Edward Hyatt electric center has six power generating turbines with a total installed capacity of 819 megawatts (MW) of electricity.

In 2017, the recording of large amounts of rainfall in the lake reception basin as well as inappropriate water evacuation maneuvers resulted in an accident that damaged the dam discharge channel and a part of the slope where the dam is set.

Dam operators were obligated to discharge water based on the chart contained in the Oroville dam reservoir regulatory manual, a flood control manual developed by the US Army's Engineer Body.

At the time of the incident, the Oroville dam reservoir regulatory manual was last updated in 1970, and the discharge diagrams were based on climatic data and drainage projections that did not take into account the current climate change or the magnitude of the significant floods previously experienced in the years 1986 and 1997 (Lake Oroville Spillway Incident: Timeline of Major Events 2017; U.S. News and World Report 2017).

After the incident occurred, it was unclear whether the obsolete operating manual or human error was significant factors in the February 2017 crisis.

8.4 Vajont Dam

On 9 October 1963, at 22:39, following maneuvers by the dam personnel, 260 million m^3 of stone slipped from Monte Toc into the dam lake. The landslide has generated an enormous wave of at least 50 million m^3 of water. Vajont dam is a double-curved, thin-arch dam, one of the largest in the world at the time, suffered no serious damage. But the flooding generated destroyed several villages in the valley and killed nearly 2000 people.

At first sight, a landslide can be considered a natural cause of an unpredictable incident and classified as a case of force majeure in the performance of the objective's activity. In fact, however, in this case, the maneuvers of lowering and increasing the water level to trigger the "controlled" slip and avoiding an accident have led to catastrophe and the abandonment of dam.

The Vajont facilities were an important economic investment, providing electricity to nearby cities and industries. After recording the first landslide, no one dared to abandon the project, seeking a controlled release option.

Research carried out after the event was more numerous. In 2018, Alan P. Dykes and Edward N. Bromhead carried out an analysis of these as a complete saying that "the logical argument is this: if the failure in 1963 was a reactivation of an old landslide, then it must have taken place in residual power, In which case a rear analysis of slope stability using the 'seat' geometry simply confirms this hypothesis in the investigator's mind.in addition, a slip of ancient land at residual power could not have accelerated and reached the observed speed and ran without another unusual mechanism having emerged on a slope to take account of this behavior" (Dykes and Bromhead 2018).

After the accident, the Italian Government insisted that the landslide was an unpredictable "act of God". Although the event was heavily affected by political decisions, in 2008, UNESCO publicly called the incident "a classic example of the consequences of the failure of engineers and geologists to understand the nature of the problem they tried to deal with" (von Hardenberg 1963; Paronuzzi et al. 2013).

8.5   Conclusions

Over the years, scientific articles have reported on the force majeure situations that various hydropower installations have experienced.

In most cases, the invocation of force majeure was generated by failures of power-generation systems. Then, there are situations where the contracted quantity of electricity cannot be guaranteed due to transmission grid failures or lack of raw materials. By developing methods for forecasting the development of energy sources and the energy market, it has been possible to avoid situations where force majeure is invoked due to mismanagement by electricity generation companies.

References

BC Hydro (2014) BC Hydro's System. Archived from the original on 2013-04-19. Retrieved 2014-07-17. "BC Hydro (July 29, 2009)". "Our Facilities". Archived from the original on 2010-07-31. Retrieved 2009-12-22. "BC Hydro (2014)". BC Hydro 2014 Annual Report (PDF). Vancouver. Retrieved 2014-07-20

Dykes AP, Bromhead EN (2018) The Vaiont landslide: re-assessment of the evidence leads to rejection of the consensus. Landslides 15:1815–1832. https://doi.org/10.1007/s10346-018-0996-y

https://vancouversun.com/business/private-power-producer-disputes-b-c-hydros-cancellation-of-20-million-in-purchases

https://wikimili.com/en/BC_Hydro

http://www.esa.int/Applications/Observing_the_Earth/Copernicus/Sentinel-1/Sentinel-1_maps_flash_floods_in_Laos

https://www.internationalrivers.org/wp-content/uploads/sites/86/2020/08/ir-factsheet-2_year_xe_pian_dam_collapse_1_0.pdf

https://www.rfa.org/english/news/laos/soil-05282019153902.html

Lake Oroville Spillway Incident: Timeline of Major Events February 4-25. http://www.water.ca.gov/orovillespillway/pdf/2017/Lake%20Oroville%20events%20timeline.pdf

Paronuzzi P, Rigo E, Bolla A (2013) Influence of filling-drawdown cycles of the Vajont reservoir on Mt Toc slope stability. Geomorphol J 191:75–93. https://doi.org/10.1016/j.geomorph.2013.03.004

U.S. News and World Report. Associated Press. 16 Feb 2017. Retrieved 23 Feb 2017

von Hardenberg WG (2011) Expecting disaster: the 1963 landslide of the Vajont Dam. Environment & Society Portal, Arcadia, no. 8. Rachel Carson Center for Environment and Society. https://doi.org/10.5282/rcc/3401

Chapter 9
Romania, in the European Energy Market

Abstract To understand the situation of Hidroelectrica Corporation and its contractual partners when it terminated the contracts concluded by invoking force majeure, we considered it useful to present a history of the evolution of the Romanian energy market and the profile of Hidroelectrica Corporation.

Keywords Energy source · Hidroelectrica Corporation · Contracts

9.1 Energy Market History

Due to some peculiarities of the energy industry, all states considered their full involvement in the energy sector as a normal and necessary action to ensure the premises of economic development.

For decades, a paradigm has been built by which the production, transport and distribution of energy belonged exclusively to the state.

The structure of this model was dictated by:

- the exclusive rights of the state to build and operate in the energy sector;
- lack of any form of competition;
- detailed regulations;
- high degree of planning and strict control;
- integrated vertical operation;
- tariffs based on production costs.

This model, generated by the political context and the degree of economic and technical development at that time, worked for a long time, accumulating in some cases the growing dissatisfaction of consumers with the fact that in none of the operating phases of the energy system, they are not part of the decision-making process, or in other situations that they are trapped in an energy system from which they cannot get out. Lack of competition or political ideologies has led in many countries to situations where energy was delivered especially to certain areas or did not cover power consumption in certain periods of time.

This rigid, traditional, government-energy relationship has been affected for some time by a seemingly irreversible change. The old certainties have begun to falter, and the unconditional acceptance of centrally made decisions no longer works, more and more obviously after the 1990s. The new wave that takes the place of centralized regulation is the one that is based on competition.

Natural monopolies, either state-owned or under its control, which operate in a technically centralized configuration, are beginning to disintegrate and reorient themselves to customers and competition. The characteristics of the new type of approach are different, namely:

- separation of activities, to allow competition whenever possible, instead of vertical integration (production, transport, distribution, maintenance);
- the freedom to invest in competitive activities (instead of centralized planning);
- the freedom to contract at competitive rates (instead of the fixed rate);
- access to network and infrastructure;
- supervision of the system by independent regulators (instead of the government);
- adaptation to information technology.

The end of Second World War was for many countries the moment when reconstruction was largely based on energy, and therefore, the energy sector must be developed and supported by the state. The industries were nationalized, and in order to avoid the abuse of power, it was resorted to the solution of public property and/or public control. This is how, among others, Electricite de France and Gaz de France were born in 1946, ENEL in 1962 in Italy and the Ministry of Electricity in Romania.

The global energy crisis of the 1970s led to energetic interventions by industrialized states in the energy sector. A new issue has appeared on the European political agenda, namely that of security of energy supply. Expensive programs have been initiated for the construction of nuclear power plants, and subsidies for alternative energy have been allocated. The International Energy Agency has been set up to monitor the allocation of financial resources and to encourage the diversification of alternative forms of energy. At the same time, national energy policies and implementation agencies have begun to appear.

However, some interventions planned in this traditional way have proved to be hasty or even useless, so the ability of governments alone to intervene in energy policy has begun to be questioned.

Advocates of the new open approach to the market have been appearing in the UK and US since the 1970s. Certain operating structures that existed in isolation, especially in the US, represented by independent producers charging energy in a public grid, raised the question of whether this type could not expand, expanding the number of players in the sector and encouraging competition, because in the future to create the free market.

In the mid-1980s, the new approach began to gain more and more followers. The taboos of state control over the energy sector have begun to fall, especially under the influence of two phenomena: the globalization of the world economy and the emergence of various government initiatives to liberalize energy markets. Globalization has brought into question the role of nation states, not reducing, but transforming their functions and depoliticizing the national space for some economic sectors. As an immediate consequence of globalization, liberalization necessarily implies a transfer of responsibility from the state to the private sector, at the same time as the appropriate takeover of regulatory powers by government agencies (Barroso 2006).

The Treaty establishing the European Coal and Steel Community (ECSC) and later the Treaty establishing the European Atomic Energy Community (Euratom) represented at European level the premises for the emergence of the European Union but first showed the need for a unitary development of the energy system. The Single European Act (1987) marked a turning point for the single market, but energy was not of particular interest because, at the time, governments were unwilling to cede some of their control over national energy monopolies, thus not opening up to the market. The 1992 Maastricht Treaty, known as the EU Treaty, made some additions to the definition of the internal energy market (IEM) concept without including a special chapter on energy (https://eur-lex.europa.eu/legal-content/EN/TXT/PDF/?uri=CELEX:11992M/TXT&from=EL). The European Commission has prepared a chapter proposal, which should, among other things, have invested it with some expertise in the field. Three countries or vehemently opposed this initiative: the United Kingdom, the Netherlands and Germany. The same fate befell another commission proposal, regarding the administration of the Energy Charter by the Energy Directorate within the EC. Discussions on energy in the European Union resumed in 1997, but Member States voted against. The European Energy Charter contains articles setting out the conditions for competition, transparency, sovereignty, taxation and the environment, as well as articles on investment protection, energy transit and the treatment of disputes. The treaty entered into force in 1998. The complexity of energy production, transport and energy consumption issues has increased greatly in recent decades, with global environmental problems, climate change and depletion of natural resources.

In addition, the European Union is facing a number of specific problems, the most serious of which is the growing dependence on imported energy resources.

Also under pressure from the commitments made under the Kyoto Protocol, the European Commission launched in 2000 the third Green Paper "Towards a European strategy for security of energy supply". The final report on the Green Energy Charter, the result of an unprecedented public debate over the last 30 years, was presented by the European Commission on 27 June 2002 (Final report on the Green Paper 2002; Directive 98/30/EC 2000; Kyoto Protocol to the United Nations Framework Convention 1997).

The Green Paper highlights the need for renewable energy sources to become an increasingly important part of the structure of energy production. Conventional energy sources with lower pollutant potential (fuel oil, natural gas, nuclear energy) are reconsidered to support the development of new energy resources through them. On the other hand, care to maintain competition on the energy market; it does not give much room for maneuver to state subsidies intended to stimulate energy producers from unconventional sources. For this reason, the European Commission considers that a minimum of harmonization in the field of subsidies is needed (Fig. 9.1).

At the level of the European Union, two categories of countries can be identified in terms of primary energy sources, in three categories: net producers (Netherlands,

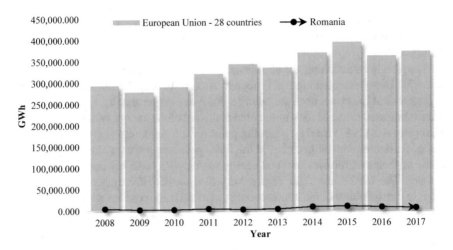

Fig. 9.1 Exports of electricity and derived heat by partner country

Denmark and Great Britain), net importers (Germany, France and Italy) and the special category of countries cohesion (Fig. 9.2).

Promoting green energy through certification or through a reform of environmental taxes is two of the most widely circulated models (Directive 96/92/EC of the European Parliament).

New and renewable sources of energy (biomass, photovoltaics (PV), wind energy, hydropower, etc.) have already become national targets for industrialized countries and not only for their energy production.

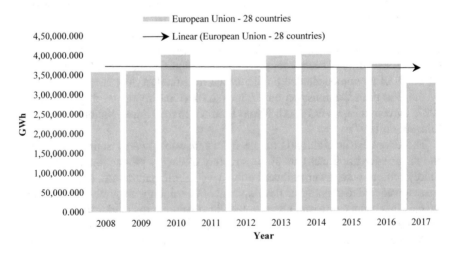

Fig. 9.2 Gross electricity production European Union (28 countries)

In Romania, engineer Henry Slade installed the first thermal power plant in Bucharest in 1882, equipped with brush-type power generators and steam boilers. The electric current transmitted through a 2 kV overhead line mounted on the pillars supplied a system with the illumination of the Royal Palace of Calea Victoriei, Cismigiu Garden and the National Theater's exterior. The thermal power plant was placed in the place of the current Central University Library (Manolea 2008). Romania's first hydropower plants have been countered since 1889—the Grozavesti hydropower plant in Bucharest is inaugurated; the current was used for public lighting and for pumping water into the mains and 1896—entering into operation Hydropower Sadu I—which supplies power to the city of Sibiu, being the first hydropower plant on the Romanian territory to use the remote transmission technology of electricity, developed by German engineer Oskar von Miller.

In the interwar period, electricity production has also increased. Many small plants have been built, serving both cities and various industrial producers. Among the most important achievements, we mention: the thermal power plant in Floresti, built in 1922–1923 to supply the Prahova oil area; the Dobroesti hydropower plant (1928–1930), created to supply electricity to Bucharest; the installation at Filaret Diesel power plant of a 5000 hp, the largest in Europe at that time; construction at the Grozavesti factory of the first hyperboloidal cooling tower of reinforced concrete; the dam of the 17-m-high hydropower plant in the Sadului Valley, built between 1939 and 1940, from rock with an inclined mask of reinforced concrete (Balan St. and Mihailescu 1985). State involvement in this field is reflected in the drafting of the Energy Law in 1924, designed to regulate electricity production, transmission and distribution and outline future projects in the field. However, in 1939, only municipalities serving 24,8% of the country's population were able to access electricity. In practice, electrification had been carried out only in cities, with villages not yet benefiting from the advantages of electricity and continuing to use traditional sources of energy.

After the end of the Second World War and the establishment of the communist regime, the concept of economic development focused on industrialization, particularly the development of heavy industry. To achieve this goal, large-scale energy resources were needed. On the other hand, electrification was a prestigious objective of communist regimes throughout Eastern Europe. Naturally, governments have given priority to the development of the energy industry and the extraction of the necessary fuels. Over 50% of industrial investment was allocated to these industries in the 1951–1955 five-year period and remained the leading investment in the next five-year period until mid-1970. New wells and mining facilities were put into operation, and improved technologies and increased employment allowed appreciable increases in production.

Romania's first great hydroenergy achievement, the Bicaz-Stejaru energy complex, was a real school for hydropower specialists who later contributed to the other major constructions on Arges, Lotru or the Danube. The building of the lake was necessary to move 2.291 households, with 18.760 inhabitants, out of 20 villages. The dam of Izvorul Muntelui—Bicaz—is 127 m high and 435 m long at the canopy; the catchment area has a total volume of 1.230 million m^3, an area of 310 ha and a length of approximately 35 km. The transport of the pressurized water from the dam lake to the Stejaru hydropower plant is carried out through a tunnel built under the Botosanu Mountain and having a length of 4655 m. The tunnel has an internal diameter of 7 m which can provide a flow of 178 m^3/s. The Stejaru hydroelectric power plant is equipped with 6 turbo units and provides an installed power of 210 MW. The Bicaz hydropower system was put into use on 1 July 1960, and on 1 October the same year, the electricity required by the Bistriței Valley is produced here.

Electricity had to be produced not only, but also distributed to industrial and domestic customers. While in the first half of the twentieth century, the idea of placing small plants as close as possible to consumers had prevailed. With the shift to power generation in high-capacity hydro and thermal power plants, the issue of building electricity transmission networks and establishing a national energy system (NES) has become increasingly serious. It is composed of electricity producers, the grid of high-voltage electricity lines, the transformer power stations and the system dispatch, the one who schedules, coordinates and supervises the activity of the NES (https://www.transelectrica.ro/web/tel/sistemul-energetic-national). After the 110 kV voltage was originally used, in the year 1960, the 220 kV voltage was chosen, and some of the networks carried the energy at higher voltages of 400 kV or even 750 kV. The first national dispatching point was installed in 1954, and a reorganization took place in 1965, following which five regional energy dispatching were set up (https://www.transelectrica.ro/web/tel/sistemul-energetic-national).

Ever-increasing electricity consumption overlapped in 70 with the oil crisis, leading to the identification of another energy alternative for Romania. The solution was investment in nuclear energy. The first concerns for the exploitation of nuclear technology for power generation date back to 1955, when talks started on acquiring a research reactor and an accelerator from the Soviet Union, which was in operation at the Institute of Atomic Physics Magurele. In 1964, discussions were held to purchase a US nuclear power plant, but the authorities preferred to postpone and give priority to large lignite power plants. In the year 1970, after a new attempt to obtain nuclear technology from the Soviet Union, cooperation with Canada was chosen. The contract for taking over the CANDU license and for designing and building a nuclear power plant in Cernavoda was concluded only in 1978, and the actual work had been considerably trended, so that in December 1989, only 45% of the investment for the first power plant group was realized. Under these circumstances, Unit No. 1 was only put into operation in 1996 and the reactor of Unit No. 2 was only inaugurated in 2007. The work on Groups 3 and 4 is at different stages of implementation (https://www.nuclearelectrica.ro/despre-noi/istoric/). The two nuclear power groups currently operating from Cernavoda account for about 17–18% of Romania's electricity consumption. The service life reached by the two groups now involves upgrading, with the first group scheduled to enter into new technology works in 2023.

After 1990, the economy and the energy industry have undergone major changes. Privatizations, tariff liberalization, bankruptcies have led to an unpredictable market with private and state actors, with some making other losses. Thus, in 1998, the nuclear activities were taken off from the Renel Electricity Autonomous Region, which was established in '90s, and the rest formed the National Electricity Company

SA (CONEL). In 2000, Conel was divided into 4 separate companies, the National Electric Energy Transport Company Transelectrica Corporation, commercial society for the production of electricity and heat Termoelectrica Corporation, commercial society for the production of electricity Hidroelectrica Corporation, commercial society for the distribution and supply of electricity Electrica Corporation. In 2002, Electrica Corporation was split into 8 electricity distribution and supply subsidiaries, and a privatization strategy was adopted for Electrica Banat and Electrica Dobrogea, resulting in the purchase of majority packages by the Italian Company Enel. Privatizations in the sector continued after 2005, when Electrica Oltenia was taken over by Czech Republic (CEZ), Electrica Moldova de EON (Germany) and Electrica Muntenia Sud, also by Enel. The privatizations were aimed, on the one hand, at bringing revenue to the state budget, on the other hand, at attracting powerful companies in Romania, capable of improving management and making investments to ensure the modernization of Romania's energy industry. They have also facilitated the process of adapting Romania to the policies and institutional mechanisms prevailing in the European Union.

The development needs of the electricity industry, as well as the assumed obligation to comply with European Union practices, have put the issue of gradual liberalization of the electricity market on the agenda. This refers to both access to, choice of economic operators and the alignment of Romanian prices to current levels on international markets (Fig. 9.3).

The national energy strategy for 2007–2020, adopted by the government in the context of its accession to the European Union and updated in 2011, states as a general objective "to meet energy needs both now and in the medium and long term at the lowest price, suitable for a modern market economy and a civilized standard of living, under quality conditions, food safety, respecting the principles of sustainable development". To this end, in line with the "Energy—Climate change"

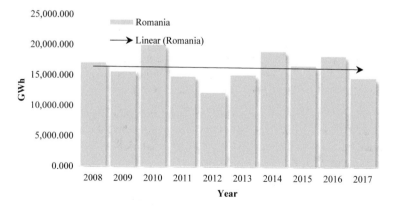

Fig. 9.3 Gross electricity production Romania. *Data source* http://www.insse.ro/cms/files/eur ostat/adse/baze%20de%20date%20eurostat.htm

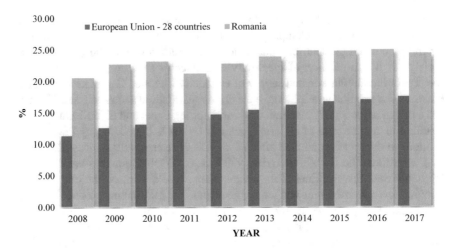

Fig. 9.4 Renewable energy sources Romania and European Union (28 countries)

package agreed by the European Council and adopted by the European Parliament in December 2008, the focus will be on:

- assuring the energy security;
- sustainable development (increasing energy efficiency; promoting renewable energy production) (Fig. 9.4);
- reducing the negative impact of the energy sector on the environment, etc.;
- increase competitiveness (development of competitive electricity, natural gas, oil, uranium, green certificates, greenhouse gas emission allowances and energy services);
- liberalization of energy transit;
- ensuring permanent and non-discriminatory access of market participants to the transmission, distribution and international interconnections;
- continuation of the restructuring and privatization process, especially on the stock exchange, within the electricity, thermal and natural gas sectors;
- continue the restructuring process for the lignite sector in order to increase profitability and access to the capital market.

According to the Romanian National Reform Program in line with the Europe 2020 Strategy, the national targets on climate change and sustainable energy are European Council Conclusions (2010), Europe 2020: A Strategy for Smart (2010), Smarter. Greener (2019):

- greenhouse gas emissions 20% lower than in 1990;
- increase by 24% the share of renewable energy in gross final energy consumption.

Romania's national energy efficiency target for 2020 is to save 10 Mtoe of primary energy, which represents a 19% reduction in the volume of primary energy consumption (52.99 tep) forecast in the PRIMES 2007 model for the realistic scenario.

Achieving this goal means that primary energy consumption will be 42.99 Mtoe in 2020, while the projected total energy consumption will be 30.32 Mtoe.

Renewable energy and carbon markets are two emerging markets and have emerged and developed gradually in the recent years. The development of renewable energy production has begun to be politically and financially supported, ensuring a competitive price for the energy produced in these ways (Ajadi et al. 2020; Berntsen et al. 2021; Liu et al. 2021).

At present, in Europe, we are witnessing an increase in energy production prices amid the restrictions created by the COVID-19 epidemic, the abandonment of the use of coal for thermal power plants and even the closure of some nuclear production units. The global energy market is now undergoing another important moment of transformation (Xu et al. 2021).

Lack of investment in efficient energy transmission systems limits the expansion of photovoltaic or wind farms in some countries. The dependence of the energy industry on fossil fuels (gas, oil) and the geostrategic movements of certain countries and organizations producing oil and gas generate major changes in the short and medium term in this industry.

Increasing the efficiency of hydropower plants, wind turbines and photovoltaic turbines is now being translated into high profits for producers, given the high selling prices of energy.

Thus, the analysis of the financial sustainability of an electricity company must be analyzed over longer periods of time and not in short sequences.

9.2 Hidroelectrica Corporation Company, Romania

One of Romania's largest electricity production companies invokes force majeure clauses in 2011 and 2012 and denounces several contracts for the supply of electricity to production or trader companies.

According to the articles of incorporation of the company, Hidroelectrica Corporation to produce and sell electricity by performing commercial acts corresponding to its object of activity in compliance with the Romanian legislation in force.

Hidroelectrica Corporation has the following as its object, under the code proposed by order No. 337/2007 (classification NACE-rev.2): Main activity: 351—"production, transmission and distribution of electricity". Main activity: 3511—"electricity production" and other secondary activities.

The registered capital of Hidroelectrica Corporation is 80.056099782% owned by the Romanian state through the Ministry of Energy and 19.943900218% by the "Property Fund".

Hidroelectrica Corporation is the largest system service provider in Romania (at national level, in 2019, it accounted for 70.97% of the required secondary adjustment band, 82.94% of the rapid reserve requirement, 100% of the reactive power service that is either cut off or absorbed from the grid into the secondary voltage regulation band), ensuring the stability of the national energy system. The hydroelectric plants have a capacity of 187 with a capacity of 6422.17 MW.

The 6422.17 MW in operation within Hidroelectrica Corporation is installed in 187 power plants and pump stations (of which CHE < 10 MW is 74 and CHE > 10 MW is 108).

The number of operating groups is 430 (of which 169 in plants with installed power < 10 MW, 250 in CHE > 10 MW) and 11 are pumping groups.

The company has a turnover of between lei 2.4 and 3.4 million (EUR 500–700 million) between 2010 and 2015, fluctuating depending on the hydrological regime of the rivers used in energy production.

The main business activities of the company are:

- electric power generation in hydroelectric power plants;
- sale of electricity;
- provision of ancillary services for the national energy system;
- provision of water management services from the company's own water reservoirs, by supplying raw water, flow adjustments, flood protection, flow rate assurance and other common services for water management;
- ensuring navigation on the Danube River by locking;
- control and mitigation of flood waves for the transit of disastrous flow rates in hydroelectric power facilities on inland rivers under the Hidroelectrica Corporation administration.

The company conducts its business under licenses granted by the National Energy Regulatory Authority (NERA), which are renewed regularly, as follows:

- Licence no. 332/2001 for the commercial operation of electric power generation capabilities, including the supply of ancillary services, updated by NERA Decision no. 509/2017, valid until 24.07.2026.
- Licence no. 932/2010 for electric power supply, updated by NERA Decision no. 768/2016, valid until 01.06.2020.

The company's organization structure includes 7 subsidiaries without legal personality, spread in the country's territory, namely: Bistrița HB—13 Locotenent Draghiescu Street, Piatra Neamț; Cluj HB—1 Taberei Street, Cluj Napoca; Curtea de Argeş HB—82-84 Basarabilor Street, Curtea de Argeş; Hateg HB—23 Nicolae Titulescu Blvd., Hateg; Porțile de Fier HB—2 I.C. Bibicescu Street, Drobeta Turnu Severin; Râmnicu Vâlcea HB—11 Decebal Street, Râmnicu Vâlcea; and Sebeş HB—9 Alunului Street, Sebeş.

In 2011, the highest prices were paid in those countries where taxes and levies contributed a considerable share of the final consumer price (e.g., Germany, Denmark), or in countries with limited or non-existent interconnections to other countries (Ireland, Cyprus and Malta). In the eastern part of Central Europe, particularly in those countries that joined the EU during the last decade, prices are generally lower than in the western part of the continent. This might be due to the widespread practice of regulated prices and to the incorporation of socio-politically motivated price subsidies in the electricity tariffs. The average wholesale energy price at Europa level was 196 euros/MWh. It varied between 86 and 110 euros/MWh in Bulgaria, Latvia, Estonia, Lithuania, Romania and 232 euros/MWh in Spain and 283 euros/MWh in Norway (https://ec.europa.eu/eurostat/statistics-explained/index.php?title=Archive: Energy_price_statistics&oldid=73524).

In the first half of 2020, the average wholesale energy price at Europe fell to around 33.5 euros/MWh. It ranges from 15.1 euros/MWh in Norway or 16.7 euros/MWh in Sweden, 27.2 euros/MWh in Luxembourg, Latvia, Lithuania and 41 euros/MWh in Poland, Hungary and Romania. The maximum price reached is 50.4 euros/MWh in Greece (https://ec.europa.eu/energy/sites/ener/files/qr_electricity_q1_2020.pdf).

In this context, press agencies announce that Hidroelectrica Corporation invoked force majeure, terminating several commercial contracts it had with different consumers or electricity suppliers.

"Romania's hydro electricity generator Hidroelectrica has declared force majeure on all its delivery contracts, slashing its August production to a historical minimum, according to a statement sent to the press on Monday. Hidroelectrica Corporation said it would cut its production to 700 GWh from the planned 1.1 TWh for August, the lowest production levels since the company was set up in 2000.

The decision will be become effective for all its 23 bilateral contracts on Tuesday, and on 10 August for the six contracts concluded as part of a tender held on the OPCOM power exchange last month.

The company said in the statement that the contractual deliveries will be reduced proportionally with the decrease in production.

"July 2012 has had the highest temperatures recorded in the last 60 years and low hydro levels in all production areas owned by Hidroelectrica SA," the company said.

"If we had not decided to call force majeure this month, we would have had to buy on the free market at prices well above those in the bilateral contracts," the company stated. Hidroelectrica insisted that it would seek to keep the force majeure period to a bare minimum.

Price spikes

The announcement coincided with a spike in the daily outturn on OPCOM, which jumped to a seven-month high of Romanian New Lei 357.88/MWh (€77.14/MWh), according to ICIS data collected during working days. Participants said the spot is driven by the hydro shortage as well as the above-seasonal temperatures across the region.

Traders have been concerned about such a move for some weeks, but September '12 Baseload was reported dealing at New Lei 237.00/MWh on Monday, also the highest traded level recorded on the contract and the highest traded level for a front-month contract since 26 January, according to ICIS data.

Prices for Calendar Year 2013 Baseload had also risen, reaching New Lei 226.00/MWh on 3 August, the highest closing level since ICIS began to assess the contract this year.

Contracts cancelled

Last month, the state-owned generator decided to cancel six long-term contracts that were renewed last year with a number of customers at prices nearly half the official market price.

The energy freed up to the market as a result amount to 6.5 TWh annually. However, the company added that its August production obligations are as high as 926 GWh. Hidroelectrica's force majeure announcement on Monday is the second in less than a year. The producer had to take a similar decision last September after it was compelled to buy electricity on the free market to honour its long-term contractual obligations.

The prolonged drought as well as the low prices charged under the long-term contracts renewed last year dried up its finances and forced it to declare insolvency" (https://www.icis.com/explore/resources/news/2012/08/06/9584447/romania-s-hidroelectrica-announces-force-majeure-on-ele ctricity-contracts/).

Romania's Hidroelectrica Corporation to increase power deliveries, *By Reuters Staff*

BUCHAREST, Nov 26 (Reuters)—Romania's insolvent state-owned hydro power producer Hidroelectrica plans to lift a force majeure activated earlier this year as of Dec. 1 and pledged to fulfil all contracts, its judicial administrator said on Monday.

The company had declared force majeure, a clause provided in contracts where buyers or sellers are allowed to renege on their commitment because of a situation that is beyond their control, after a severe drought slashed its output.

"We will halt the clause and buyers will get contracted power quantities," Remus Borza, Hidroelectrica's judicial administrator, was quoted as saying by state news agency Agerpres.

Borza said production is estimated to be below 12 TW this year, while the company is expected to record losses of about 70–80 million euros ($103.67 million). He sees output rising to 13 TW in 2013.

Hidroelectrica SA, which has an installed capacity of 6400 megawatts and is Romania's cheapest power producer, was declared insolvent in June. ($1 = 0.7717 euros) (Reporting by Ioana Patran; editing by Radu Marinas and Jason Neely) (https://fr.reuters.com/article/romania-hidroelectrica-idINL5 E8MQ2D120121126).

Daily News

Hidroelectrica calls force majeure clause in energy supply contracts because of drought Romania's biggest energy producer, Hidroelectrica, has activated the force majeure clause on its energy supply contracts because of the drought, according to Ziarul Financiar newspaper. The energy producer can no longer perform its contractual obligations due to the low level of rain.

According to the Romanian National Institute the final electricity consumption in the first seven months rose by 4% to 30.46 billion kWh in Romania, compared to the same period in 2010, while electricity sources were with 7.1% higher, to 36.47 billion kWh. The public lighting consumption decreased by 14.8%, while household consumption increased by 3%.

The lack of rain in the last period has also triggered the Danube's lower water. Earlier this month, the river's level decreased to depths of 1.2–1.4 m in some areas, so navigation was blocked because of some critical areas.

Romania: Hidroelectrica won court cases against electricity traders for force majeure clause 25. February 2015./SEE Energy News.

Bucharest Court rejected the appeal filed by Elsid Titu SA which called on force majeure clause in electricity supply contract concluded with Hidroelectrica SA. Elsid court ordered to pay costs in the amount of about 29,500 lei (6620 euro).

Elsid, energy trader and manufacturer of electrodes filed a case against Euro INSOL manager of Hidroelectrica for force majeure clause for the period August 1, 2012–December 1, 2012, the clause which had the effect of reducing the amount of energy that Hidroelectrica delivered by a Elsid the period.

Representatives of Hidroelectrica stated that in 2012, the company has experienced the worst drought in its history, the lowest recorded Danube flows in the last 150 years. However, 2012 has recorded the lowest ever recorded Hidroelectrica production (11.8 TWh). Since the company had contracted obligations 20 TWh in that year and the market price was around 250 lei/MWh, Hidroelectrica was forced to resort to this measure to reduce losses.

At the opening of insolvency proceedings, Elsid have a fixed contract price of 103 lei/MWh (23.11 Euro/MWh), which did not even cover the cost price of 171 lei/MWh (38.37 Euro/MWh), which was much lower than the market price. The contract between Hidroelectrica and Titu Elsid generated a loss of 97 million lei between 2006-31, 2012.

11 traders have concluded that Hidroelectrica electricity supply contracts, Elsid, Alro and Electrocarbon accepted the conditions for increase the price. If Elsid price increased from 103 lei/MWh to

180 lei/MWh (40.39 Euro/MWh), energy trader has to pay retroactively the difference between the contract price and the new price negotiated (https://serbia-energy.eu/romania-hidroelectrica-won-court-cases-aga inst-electricity-traders-for-force-majeure-clause/).

9.3   The "Force Majeure" Mentioned in Contracts Concluded by Hidroelectrica Corporation

Hidroelectrica Corporation concluded a number of contracts with several major Romanian energy consumers or energy traders. The contract itself, written on six pages, was accompanied by a number of 4 annexes, which constitutes another seven pages. A number of amendments are subsequently entered into which supplement the contract initially.

Thus, in the contract concluded between SC Hidroelectrica Corporation as supplier and ALRO Corporation as eligible customer, contracts are concluded for the supply of electricity to eligible customer No. 47/02.09.2005; a force majeure Article is mentioned. This article is as follows: "force majeure".

Article 11.

(1)   the parties shall be exempt from all liability for failure to fulfill all or part of the obligations arising under this contract if this is the result of force majeure. Circumstances of force majeure may arise in the course of performance of this contract following extraordinary events that could not be taken into account by the parties to the contract and which are reasonably beyond the will and control of the parties.

(2)   the party invoking force majeure must notify the other parties in writing within 48 h of its occurrence, with confirmation of the applicant body at the place of the event constituting force majeure.

(3)   failure to comply with the obligation to communicate force majeure does not affect its discharge of liability, but requires the party invoking it to make good any damage caused to the Contracting Party by the failure to communicate.

On 8.09.2005, i.e., 5 days after the date of conclusion of the original contract No. 47 EC, the first amendment is concluded. Article 11 is amended as follows:

Article 11—force majeure.

11.1 Sections 11(1) and 11(2) of the contracts shall be replaced by the following:

(1)   if, as a result of an event of force majeure, a party is unable to fulfill, in whole or in part, its obligations under this contract (other than obligations to pay sums of money), The party concerned must notify the other parties of the event(s) of force majeure and the obligations affected within 48 h of the occurrence of the force majeure event(s), in writing, As well as provide full details of all relevant

situations and confirmation by the competent authority of the territorial area in which the force majeure event occurred and will continue to keep the other Party informed.

(2) by submitting the notification specified in Section 11(1), the parties' obligations will be suspended, but only to the extent and for as long as the fulfillment of the obligations is indeed affected by force majeure.

After Section 11(3) of the contract, the following new sections shall be added:

(1) any period or time limits referred to in this contract within which a particular obligation must be fulfilled shall be extended by a period of time equal to the time of delay caused by the event of force majeure.

(2) the parties shall endeavor to remedy the effects of force majeure as soon as possible.

(3) the period of force majeure shall end when the party submitting the notification under Section 11(1) sends a new notification stating reasonably that it is capable of refulfilling all its obligations under this contract and resuming all the obligations referred to therein in that notification.

(4) if the period of force majeure lasts 12 months, either party shall be entitled to withdraw from this contract unilaterally by giving prior written notice within 60 (sixty) days, after which the provisions of Section 9(2) shall apply. These provisions are established provided that always, Where the damage or deficiencies caused by force majeure are such that their remedy would last longer than the 12-month period referred to in this Section 11(7) and the seller or the law buyer to remedy the damage or deficiencies and propose a plan Remedy to be followed diligently by the seller or buyer, then the right to terminate the contract due to force majeure shall be suspended for the period during which the seller or buyer implements this plan.

Appendix 1 of this additional agreement provides the definition of force majeure as follows: "Any event or combination of events causing or having the effect of an infringement, A delay in, or impediment to, the performance of the obligations of a party to this contract; and (i) is beyond the control of that party; and (ii) shall not be attributed to any deliberate or unlawful act or omission of that party or to an infringement by a Party Any obligation under this contract which, subject to such qualifications, will include: Strikes, closure of the enterprise, other industrial disruptions, government actions, floods, hurricanes, earthquakes, natural disasters, nuclear radiation, rebellions, riots, revolutions, civil wars, insurrections, acts of terrorism or war".

No other force majeure provisions are provided for in the next 17 amendments concluded and their annexes.

The argument that the contract was terminated between the parties was based on the Article of force majeure, due to a decrease in electricity production based on the installation of a dry period with reduced water flows.

The contract and the amendments concluded that the natural events constituting force majeure are only "…, floods, hurricanes, earthquakes, natural disasters, nuclear radiation, …".

Looking at these types of extreme phenomena invoked in the contract, we notice some inaccuracies with reality. Flooding is often caused by heavy rainfall, rapid snow melting or a storm from a tropical cyclone or tsunami in coastal areas. There are 3 common types of flooding:

- Flash floods are caused by rapid and excessive rainfall that raises water heights quickly, and rivers, streams, channels or roads may be overtaken.
- River floods are caused when consistent rain or snow melt forces a river to exceed capacity.
- Coastal floods are caused by storm surges associated with tropical cyclones and tsunami.

In the case of hydroenergy developments, floods can cause damage to dams by exceeding the dam crown, by erosion of the dam foundation, damaging the overflow field or the drainage channels of the water discharged at the maximum flow rates. Also, during the flash flood, large amounts of alluvial deposits and flows can plug the outlets of the water-bringing galleries into the hydropower plant units.

A tropical cyclone is a rapidly rotating storm system characterized by a low-pressure center, a closed low-level atmospheric circulation, strong winds that produce heavy rain. Depending on its location and strength, a cyclone is referred to by different names, including hurricane, typhoon, tropical storm, cyclonic storm or tropical depression. A hurricane is a tropical cyclone that occurs in the Atlantic and North-eastern Pacific Ocean. The typhoon occurs in the Northwestern Pacific Ocean; in the South Pacific or Indian Ocean, considered comparable storms are referred to simply as "tropical cyclones" or "severe cyclonic storms". Romania, it is situated in the northern hemisphere at the intersection of the parallel of latitude 45° north and the meridian of longitude 25° east, is not in any geographic area where such an extreme natural phenomenon could occur.

Earthquakes occur when two tectonic plates slide together and their edges are intertwined. This gives rise to a very strong tension until one of the plates breaks and gives rise to a sudden movement.

According to the legislation for the design, execution and evaluation of the hydrotechnical works on the crossed-line (NP 076-2002, indicative NP 076-2013), the main elements to be introduced in the design of the structures are the seismic site factor, the acceleration of land ag and the time of the control period (corner) of the elastic response spectrum (T_C) (Normativ de Proiectare, Execuţie si evaluare la Acţiuni Seismice Alucrărilor Hidrotehnice din Frontul Barat). Romania has maps for the distribution of these parameters, and the data contained in it are used for all the structures designed (Fig. 9.5). Obviously, there are cases when the dam structure resisted strong seismic waves, but the basin verses slipped and clogged the lake cuvette by reducing the useful volume of water or clogging the river bed.

The term natural disasters are, in fact, extreme natural phenomena such as snow avalanches, electric discharges, tsunami, tornadoes, volcanic eruption, earthquakes, floods and hurricanes. Practical is a general expression of rare natural phenomena, partially explained above. Obviously, electrical discharges can cause damage to some

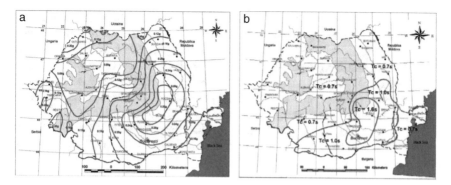

Fig. 9.5 Distribution land acceleration a_g with IMR $=$ 225 years (**a**) and corner time T_C (**b**) (Normativ de Proiectare, Execuţie si evaluare la Acţiuni Seismice Alucrărilor Hidrotehnice din Frontul Barat)

installations or parts of installations for the production and transmission of electricity. Snow avalanches affect at most some access paths, producing only short-time (1–3 months) jams. Tsunamis are phenomena that do not affect Romania's territory, especially the locations where energy dams and hydroelectric plants are located. The Black Sea, with which Romania borders, is a sea with a relatively small area and slightly inclined underwater slopes. The volcanic activity in Romania is very low, so the likelihood of volcanic eruptions affecting activity is again very low.

Therefore, the provisions of force majeure provided for in the contract and the amendments concluded between the parties are very vague. As proved during the proceedings, the provisions did not cover or mention the phenomena and their size to which the electricity producer cannot produce and deliver the contracted quantities.

9.4 Conclusions

The electricity industry in Romania has evolved over time according to the requirements of the internal market as well as the codifications generated by the accession to the European single market. The electricity generation portfolio has evolved as well as the legislation on electricity generation and marketing.

In these transit conditions, there have been periods of legislative vacuum or ambiguity, leading to situations where some energy producers and distributors have made huge profits or reached financial crisis situations, close to bankruptcy.

Hidroelectrica is the largest electricity generation company in Romania. The most important activity is the provision of system services. At the national level, in 2021, it provided about 58.05% of the required secondary regulation band, 79.07% of the fast rising tertiary reserve requirement, 26.26% of the fast falling tertiary reserve requirement and 100% of the service of ensuring the reactive energy flowed or absorbed from the grid in the secondary voltage regulation band, ensuring the stability of the

National Energy System (https://www.hidroelectrica.ro/article/20). With an installed capacity of 6372.17 MW in 187 power plants and pumping stations (of which 74 have a Hidroelectric Capacity (HEC) of less than 10 MW and 108 have a HEC of more than 10 MW), it also has the lowest electricity production costs (https://www.hidroe lectrica.ro/article/20). The management system of the company has been imposed or influenced in certain periods by political power. The lack of a coherent vision of the company's role in the energy market and the problems caused by global climate change have also created the conditions for disastrous financial situations. In this context, there have been borderline situations in which the company has experienced force majeure situations in its operations.

References

Ajadi T, Boyle R, Strahan D, Kimmel M, Collins B, Cheung A, Becker L (2020) Global trends in renewable energy investment 2020. Frankfurt School-UNEP Centre/BNEF, Frankfurt, Germany

Balan St., Mihailescu N (1985) The history of science and technology in Romania. Book House Academiei RSR, 487

Barroso JM (2006) The EU and energy: looking to the future. EU Focus September 1–3

Berntsen J, Fjellheim H, Maria K, Cathy L, Rihel AS, Zelljadt E (2021) Carbon market year in review 2020. Refinitiv, London, UK

Directive 96/92/EC of the European Parliament and the Council concerning common rules for the internal market in electricity

Directive 98/30/EC of the European Parliament and the council concerning common rules for the internal market in gas Green Paper "towards a European strategy for the security of energy supply", COM/2000/0769 final, Nov 2000

Europe 2020: A Strategy for Smart (2010) Sustainable and inclusive growth, European Commission. Publications Office of the European Union, Luxembourg. Available online: https://eur-lex.europa. eu/legal-content/EN/ALL/?uri=celex:52010DC2020

European Council Conclusions, EUCO 13/10, CO EUR 9, CONCL 2, 17 June 2010, Brussels, 2010. Available online: https://ec.europa.eu/eu2020/pdf/council_conclusion_17_june_en.pdf

Final report on the Green Paper (2002) "towards a European strategy for the security of energy supply", COM, 321 final, Iulie 2002

https://ec.europa.eu/energy/sites/ener/files/qr_electricity_q1_2020.pdf

https://ec.europa.eu/eurostat/statistics-explained/index.php?title=Archive:Energy_price_statis tics&oldid=73524

https://eur-lex.europa.eu/legal-content/EN/TXT/PDF/?uri=CELEX:11992M/TXT&from=EL

https://fr.reuters.com/article/romania-hidroelectrica-idINL5E8MQ2D120121126

https://serbia-energy.eu/romania-hidroelectrica-won-court-cases-against-electricity-traders-for-force-majeure-clause/

https://www.hidroelectrica.ro/article/20

https://www.icis.com/explore/resources/news/2012/08/06/9584447/romania-s-hidroelectrica-ann ounces-force-majeure-on-electricity-contracts/

http://www.insse.ro/cms/files/eurostat/adse/baze%20de%20date%20eurostat.htm

https://www.nuclearelectrica.ro/despre-noi/istoric/

https://www.transelectrica.ro/web/tel/sistemul-energetic-national

Kyoto Protocol to the United Nations Framework Convention on Climate Change; adopted at COP3 in Kyoto, Japan, on 11 Dec 1997

Liu Y, Yang X, Wang M (2021) Global transmission of returns among financial, traditional energy, renewable energy and carbon markets: new evidence. Energies 14:7286. https://doi.org/10.3390/en14217286

Manolea G (2008) Inventions and their histories. ALMA Book house Craiova. ISBN 978-973-1792-55-2

Normativ de Proiectare, Execuţie si evaluare la Acţiuni Seismice Alucrărilor Hidrotehnice din Frontul Barat—Revizuire NP 076-2002. Indicativ NP 076-2012

Smarter. Greener (2019) More inclusive? Indicators to support the Europe 2020 Strategy. Publications Office of the European Union, Luxembourg. Available online: https://ec.europa.eu/eurostat/web/products-statistical-books/-/KS-04-19-559

Xu Y, Ahokangas P, Yrjölä S, Koivumäki, (2021) The fifth archetype of electricity market: the blockchain marketplace. Wireless Netw 27:4247–4263. https://doi.org/10.1007/s11276-019-02065-9

Chapter 10
Settlement of the Dispute: Case of Hidroelectrica Corporation Invoking the Force Majeure Clause

Abstract This last chapter presents a technical aspect from a hydrological point of view in the conduct of the dispute in the company Hidroelectrica Corporation, which denounced the electricity supply contracts to various traders or economic agents. It is desired to present the issues invoked as being of a force majeure nature on the basis of which decisions were taken in this dispute. They are presented as they were sustained before the magistrates of the Bucharest Court, Romania.

Keywords Dispute · Bucharest Court · Contract termination

Abbreviations

EU	European Union
M/S	Member State of the European Union
ECSC	Treaty establishing the European Coal and Steel Community
Euratom	Treaty establishing the European Atomic Energy Community
IEP	The internal energy market
CEP	Common energy policy
Tep	Tons of oil equivalent
	Gross inland consumption = Primary production + Primary product receipt + Other sources (recovered products) + Recycled products + Imports + Stock changes − Exports − Bunkers − Direct use
	Available for final consumption = Gross inland consumption − Transformation input + Transformation output + Exchanges, transfers and returns − Consumption of the energy branch − Distribution losses
	Statistical difference = Available for final consumption − Final non-energy consumption − Final energy consumption
NIHWM	National Institute of Hydrology and Water Management
HS	Hydrological station

© The Author(s), under exclusive license to Springer Nature Switzerland AG 2023
D. C. Diaconu, *Force Majeure in the Hydropower Industry*,
https://doi.org/10.1007/978-3-031-27402-2_10

In order to settle the case, the Romanian court ordered, among other things, the carrying out of the hydrological expertise for which a number of 10 objectives were set, of which 4 objectives to be intimated and 6 objectives by the impugner.

With a view to establishing the real facts concerning the matters covered by the objectives of the court, I have carried out the examination of the contents of the hydrological expert report drawn up and submitted to the case file by the expert appointed by the court, examining the findings and conclusions of this report drawn up on the basis of the documents submitted for examination in the light of the evidence relied on by the parties in conjunction with the subject matter of the proceedings brought before them.

Following the analysis carried out in accordance with the principles and working methodology specific to the hydrological expertise activity, we found that there were obvious discrepancies between the actual factual situation as evidenced by the content of the expert documents and the conclusions of the expert appointed. In my opinion, these inconsistencies lead to the distortion of the truth and consequently to an unfair solution being given in the case inferred from the judgment.

In view of the fact that, as I said in the previous chapter, the hydrologist expert appointed by the court did not collaborate in the drafting of his report, my findings and my views contrary to what he established, I felt that my reasoned opinion should be brought to the attention of the honorable court.

10.1   Objectives Proposed by the Objectors

1. To establish whether the Hidroelectrica Corporation, a hydropower electricity producer and supplier, has carried out or has used reports by the relevant authorities on hydrological forecasting for the year 2011. Considering the period for which force majeure was activated in 2011 and from 30 September 2011 to 30 April 2012, respectively, statistical analyses and hydrological forecasts and during which period these forecasts were made will be taken into account.

The expert appointed for this purpose concludes as follows:

"Water forecasts are often addressed to short-term (floods) or longer-term (seasonal) events. The very long ones, for example 12 months, have a high degree of error (30–40%). The accuracy and timeliness of hydrological forecasts depend on the reliability and quantity of hydrological and meteorological information, the response time of the river basin, the rate at which the status of the basin can be assessed at any given time, the forecasting techniques or models used, the speed with which the beneficiaries receive the forecast, etc. (Diaconu and Jude 2009). Regardless of the forecasted element type, the issued forecasts imply a degree of error that is expressed in % of the real value or in absolute values (m^3/s, m^3, days, months, etc.).

Forecasts for the first month are within ± 10%. Deviations of more than 50% are recorded for the second and third months. It is undeniable that the forecast for

2011 starts with major deviations from the estimates made in September 2010 (with extremely high electricity production due to rich rainfall).

The 2011 forecast, over a three-month period, shows that deviations from the 2010 forecast have occurred since August 2011. Measures to establish a major force were taken (30.09–31.12.2011) when the deviations surpassed by minus 100 GWh, although the achievement was well below the forecast from April 2011 onwards. The forecast also imposed the force major over the next three months (October, November, December) which differed from the long-term forecast (1 year):—32% for October;—39% for November;—47% for December. The deficit over a long period of time is passed on in the coming months, mostly with a negative impact as lack of water leads to the need for accumulation. In this case, many days are required to accumulate enough water to produce electricity. This depends on a number of factors: The size of the river basin, the size of the accumulation, the way water is used, etc.".

In my opinion, the expert-designate has only partially responded to the objective. The reasons for this opinion consist of the following arguments:

The hydrological forecasts used by Hidroelectrica Corporation are provided by the National Institute of Hydrology and Water Management (NIHWM). These hydrological forecasts, accessible to the public and other commercial companies, are short term, medium term and long term, both for inland rivers and for the Danube River (www.inhga.ro/web/guest/prognoza_medie_durata_rauri).

Short-term hydrological forecasts represent a prediction of the water level (H centimeters) and the average flow rate (Q mc/s) over a period of 1 day, as well as of the trend of these parameters (decrease, increase, standstill). The short-term forecast bulletin provides information in tabular and graphic form. The forecast is designed for 50 hydrometric stations located on 39 rivers within the country. The water bulletin issued on a daily basis also meets diagnosis for the next 7 days (level and flow), for the main rivers in the country and for the Danube River, which is practically a long-term average forecast.

A brief analysis of the forecasts issued and presented in the daily hydrological bulletin issued by NIHWM between 30.09.2011 and 30.04.2012 (which were made available to Hidroelectrica Corporation at that time) can be found:

- The average expected flow rates at one-day interval are within ± 92–96% of the value of the actual recorded flow rates.
- For a two-day period, the accuracy is reduced to ± 77%.
- So with the increase in the time range, the accuracy is decreasing more and more, with the advanced value of NIHWM to attest the issued forecasts being ± 60%.

The forecast of monthly and seasonal average flow rates (1–3 months) represents the water forecast over a long period of time. Due to their lack of precision, the public forecast service did not provide any other hydrological forecasts covering a period of 1 year. Hidroelectrica Corporation forecasts the stock of next year at the level of 80–90% of the stock of the normal year for inland rivers and 90% for the Danube, with corresponding monthly values.

Thus, forecasts issued for a maximum of 3 months are determined for 66 sections located on 37 inland rivers and for the Danube at the country's entrance—Baziaș.

For the period 30.09.2011–30.04.2012, the average monthly flow is expected to be met in forecast bulletins issued 3 months in advance, i.e., from 30.06.2011. Hydrological forecasts shall be based on purely hydrological data, hydrological patterns of liquid leakage events, meteorological data and meteorological forecasts.

For forecasts with a longer time horizon (e.g., 1 year), hydrological forecasts can only be based on strict hydrological data to a small extent, as the degree of compliance with that forecast would be very small. Under these conditions, in order to increase the degree of accuracy, it is necessary to call upon the weather forecasts issued for the same time interval.

Thus, carrying out an analysis of a material issued and presented by Director of the National Meteorology Administration (N.M.A.) on 22 March 2015 with the title "observed and future climate change", it can be observed that the average air temperature at the global level, in the period 1850–2014, there is one increasing, with some absolutely normal oscillations, at the level of data processed by several international institutes (Adler and Chelcea 2014; Adler 1994; Beran and Rodier 1985; Laaha 2002; Li et al. 1994; Chilikova-Lubomirova 2014b).

The trend toward an increase in the average global air temperature is predictable and certain, with all the implications of this phenomenon: melting of the ice cap, desertification, high evaporation, etc.

In Romania since 1985, the average annual air temperature has been constantly increasing. There is an increase in the average air temperature by 0.5 °C in the range 1981–2014 to 1961–1990. In this time frame, 2012 and 2014 are characterized as part of the top five warmest years.

Looking at the evolution of annual rainfall amounts in Romania between 1961–2014, it is noted that 2011 has a country average rainfall value of 493.20 L/m², compared to a multiannual average value of 637.8 L/m². The year 2014, even if warm, has an average country rainfall of 807.7 L/m², much higher than the average.

From the analysis of the documents provided by Hidroelectrica Corporation in its 2011 estimate of production at 14.09.2010, the estimates made in relation to what was subsequently recorded can be observed (Table 10.1) (Chilikova-Lubomirova 2010, 2014a, 2015a).

Analysis of monthly energy production shows up to + 59.0% (January) or up to − 49.0% (December) compared to forecast. However, this year's average is only − 8.0% lower than the forecast, which is not a very drastic decrease that would lead to force majeure.

Apparently inexplicable are the forecasts of the Hidroelectrica's production forecast for November (1229 GWh) and December (1229 GWh), as months are in the cold season, when a large part of the volumes of water are blocked on the verts in the form of snow or ice (Fig. 10.1). The forecast values over these months show higher values than February when positive air temperatures normally lead to melting of the snow layer in the mountain area, and in the hill and plateau areas, we already have abundant liquid precipitation.

Table 10.1 Energy production forecast for 2011 that achieved (GWh)

Month	J	F	M	A	M	I	I	A	S	O	N	D	Total
Forecast quantity (GWh)	1150	1063	1377	1668	1831	1637	1404	1155	1000	1015	1105	1229	15.640
Quantity realized (GWh)	1830	1492	1542	1599	1477	1187	1237	1222	887	676	660	634	14.443
Deviation (±%)	+ 59.0	+ 40.0	+ 12.0	− 5.0	− 20.0	− 28.0	− 12.0	+ 6.0	− 12.0	− 34.0	− 41.0	− 49.0	− **8.0**

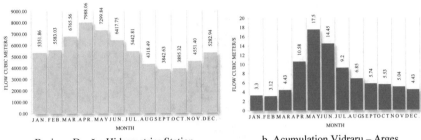

a. Baziaş – Dunăre Hidrometrics Station

b. Acumulation Vidraru – Argeş

Fig. 10.1 Multiannual average flow hydrograph

In the long term, there are models that can predict the evolution of hydrometeorological parameters on a monthly basis for the Romanian territory. These models provide data at monthly level for representative points of the territory. There is a trend toward declining water quantities on river basins, especially during the hot season of the year, and an increase in the rest. In the forecast of the monthly quantity produced by Hidroelectrica Corporation, these data should have been included in order to make the correct contracting/operating decisions.

2. To determine the rivers and lakes in Romania which directly affect hydrologically the level of electricity production for Hidroelectrica Corporation and whether their evolution has been taken into account on the basis of the hydrological forecast reports referred to in objective

The conclusion of the expert (Dr. Gheorghe Romanescu) appointed to that objective shall be as follows:

- The Danube basin with hydroelectric power stations Iron Gates I and Iron Gates II;
- Olt basin—the whole of the Olt River, including the Lotru subbasin with the Vidra dam and the Ciunget-related power plant;
- The Arges basin with the Corbeni ending hydroelectric power plant of Vidraru Lake;
- Bistrita basin with the end hydroelectric power plant of Lake Stejaru;
- Râul Mare basin with the end Retezat hydroelectric power plant of Gura Apelor lake;
- The Sebeş basin with the lakes Oaşa (Gâlceag hydropower plant), Tau (Sugag hydropower plant);
- The Somes basin with the Fantanele Lake and the Mariselu associated central;
- The Criş basin with the Drăgan Lake and the Remeti hydroelectric power plant;
- The Cerna-Motru-Tismana basin on the Cerna River with the hydroelectric power plants Motru and Tismana subterranan;
- The Bistra Marului Basin, the Ruieni hydroelectric power plant".

Table 10.2 Share of project energy produced in 2011 and 2012

No.	River basin	Share of project energy produced (%)	Share of project energy produced in the year 2011 (%)	Share of project energy produced in 2012 (%)
1	Bistriţa (including Prut and Siret)	10.37	11.16	7.91
2	Buzău	2.35	3.05	2.67
3	Someş	3.39	3.24	3.19
4	Argeş	6.70	6.62	5.34
5	Raul Mare	5.15	2.84	2.81
6	Crişul Repede	2.60	2.26	2.18
7	Dunăre	37.66	39.16	51.23
8	Olt	23.43	25.31	18.98
9	Sebeş	3.49	3.46	2.37
10	Cerna (including Motru and Jiu)	2.70	1.91	2.02
11	Bistra Mărului	2.16	0.99	1.30
	Total	100	100	100

Data source Chilikova-Lubomirova (2015a)

In my opinion, the expert-designate has only partially responded to the target by indicating the river basins, according to the data provided by Hidroelectrica Corporation, without also motivating the influence of the hydroelectric production units within them and the share of energy in the total production produced. In the statement of reasons for this opinion, the following points are made.

According to the data provided by the operation service of Hidroelectrica Corporation, the project energy share of the average hydrological year is thus produced (Table 10.2).

Compared to the share of project energy production in the average hydrological year (%), there are some variations in production in 2011 and 2012, respectively, some hydropower plants produced a greater part of the total energy than normal, and others produced a smaller or similar proportion of the total energy.

In the objective analysis of the sources and share of electricity production produced by Hidroelectrica Corporation, account shall be taken both of the capacities (useful volumes) of the reservoirs and of the water exploitation level of them and of the installed power of hydropower groups (Table 10.3). The production capacity at the 2011 level of the branches is given by:

- 140 microhydro plants with installed power below 4 MW with a total of 287 hydropower groups totaling 111.86 MW;
- 23 hydroelectric plants with installed power between 4 and 10 MW with a total of 46 hydropower groups totaling an installed capacity of 165.68 MW;

Table 10.3 Main hydropower lakes operated by Hidroelectrica Corporation

Characteristics of accumulation lakes						
No.	Accumulation lake	Min. level [mdM]	Max. level [mdM]	Useful volume [mil m^3]	Dead volume [mil m^3]	Energy project GWh/year
1	Vidraru	740	830	409.52	41.1	400.00
2	Izvorul Muntelui	470	513	918.36	203.51	434.500
3	Vidra	1237	1289	298.3	42	1100.00
4	Fântânele	941	991	202.23	10.74	390.00
5	Drăgan	792	851	103.91	8.04	200.00
6	Oașa	1210	1255	114.12	11.98	260.00
7	Tau	734	795.9	26.46	–	260.00
8	Valea lui Iovan	611.95	685	116.3	3.7	130.00
9	Gura Apelor	974.5	1060	200.54	9.99	605.00
10	Siriu	523	560	46.69	12.8	122.00
11	Poiana Mărului	555	600	41.8	5.63	264.00
12	Pecineagu	1047	1113	60.27	3	57.00
13	Rîușor	843.6	907	51.1	2.4	43.80
14	Cerna	212	237	14.68	1.5	25.00
15	Stânca Costești	78	90.8	557.68	180	65.00

- 106 hydroelectric plants with installed power exceeding 10 MW with a total of 247 hydropower groups totaling an installed power of 6074.27 MW;
- 5 pump stations which add up to an installed capacity of 91.5 MW.

By linking these data to the value of the total production in 2011 and 2012, we can estimate with an approximation the quantity produced by the hydropower units in the portfolio (Fig. 10.2).

Of the financial statements made available by Hidroelectrica Corporation publicly available during 2014, total electricity production in 2010 was 19.852 GWh; in 2011, it was 14.710 GWh; in 2012, it was 12.065 GWh; and in 2013, it was 14.823 GWh. Analyzing quantitatively the electricity production generated by the water of the Danube River, we find that out of the total electricity produced by Hidroelectrica Corporation in 2011, it accounted for about 576.043 GWh and in 2012 about 618.089 GWh, which is about 7% more than in the previous year.

The variability of energy production on river basins is, however, not only due to the variability of multiannual leakage from these basins, but also due to technological reasons such as the re-engineering of hydroelectric groups, technical maintenance, the clogging of the reservoirs and the reduction of their useful volume, but also the poor management of the accumulation in the sense of the lack of vision over long

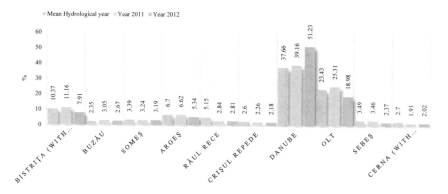

Fig. 10.2 Evolution of the share of energy produced on hydrographic bases by Hidroelectrica Corporation (Series 1—mean hydrological year; Series 2—year 2011; Series 3—year 2012)

periods of the hydrological regime, which results in poor regulation of water flows flowing from the reservoirs.

3. Determine the multiannual, monthly average values considered normal, which are the values considered as water resource scarcity that impose restrictions on the part of the electricity producers and the levels at which the situation can be qualified as "natural disaster".

The expert appointed for this purpose concludes that: "The average flow rate in the years 2011 and 2012 was lower than the average annual flow rate for all rivers equipped for the production of electricity. The lowest values have been found on the Lotru, Sebeş, Cerna, Bistra, Târgului and Prut rivers where values reach 50% of the average annual flow. The other rivers account for 60–70% of the average annual flow rate. The values for both years are relatively close. The Danube, the largest producer of hydropower, is worth 76.38% in 2011 and 79.28% in 2012" (Table 10.4).

In my opinion, this conclusion can be complemented by the following point. For several inland rivers and the Danube River, the values of the flows considered in the normal hydrological range have been calculated (Table 10.5).

Hidroelectrica Corporation did not establish by contract or other amendments a minimum flow of liquid leakage at which it can no longer honor its obligations to deliver energy to its customers, or the flow necessary to announce a natural disaster. Natural disaster or force majeure cannot be the case because the hydrological regime recorded on rivers in 2011 and 2012 has varied in normal flow rates over a period of at least 40–50 years. Thus, it is only on the basis of historical flow analysis that the minimum and maximum variations of water flows flowing on the inland rivers and on the Danube can be determined (Diaconu and Şerban 1994; Mătreaţă 1997).

We understand the invocation of force majeure or natural disaster, when a river that has never dried since being scientifically recorded suddenly dries up, or when an earthquake or a landslide affects dam and associated energy constructions, and there is no possibility of generating electricity or other phenomena of this kind which are completely unpredictable and independent of the producer's will.

Table 10.4 Comparison of the average flow rate of the years 2011 and 2012 with the average flow rate of the annual flow rate for rivers equipped for the production of electricity

River/section	Year/water flow m³/s													Average flow (m³/s)	Average flow 2011 and 2012 (%)
	2000	2001	2002	2003	2004	2005	2006	2007	2008	2009	2010	2011	2012		
Argeș/Vidraru	13.52	17.21	14.77	13.90	20.20	25.28	19.56	20.18	17.62	18.69	24.74	13.58	11.73	19.76	68.72/59.36
Bistrița/Iz. Muntelui	38.87	48.51	61.82	30.48	44.63	59.96	66.23	49.86	54.88	34.85	74.36	38.58	32.71	48.36	79.77/67.63
Lotru/Vifra	10.06	13.40	11.70	12.94	15.27	21.00	18.46	18.50	14.14	17.37	22.01	10.42	10.17	19.59	53.19/51.91
Someș/Fântânele	10.21	15.73	13.64	8.57	14.42	13.96	16.26	11.92	11.08	11.98	16.74	7.85	8.69	11.97	65.58/72.59
Drăgan/Drăgan	6.13	9.99	8.22	5.29	9.41	9.45	9.70	7.45	7.62	7.54	10.39	4.68	6.05	8.10	57.77/74.69
Sebeș/Oașa	3.80	4.51	4.43	4.23	5.61	7.84	6.55	7.87	6.81	6.27	7.87	5.37	4.29	8.28	64.85/51.81
Cerna/V lui Iovan	6.08	7.33	6.46	8.30	10.94	11.76	10.59	9.07	8.16	10.44	12.64	4.89	6.17	10.31	47.42/59.84
Râul Mare/Gura Apelor	8.19	9.87	8.31	8.09	12.07	12.68	12.10	12.55	10.30	12.79	14.65	6.76	7.66	10.37	65.18/73.86
Buzău/Siriu	6.74	7.30	8.45	4.76	7.53	19.02	10.15	8.51	6.87	7.89	13.18	7.30	7.42	9.80	74.48/75.71
Bistra/Poiana Mărului	5.78	7.56	5.79	4.46	6.56	8.69	8.55	6.81	5.89	5.53	7.40	3.64	4.80	7.24	50.27/66.29
Dâmbovița/Pecineagu	2.39	2.83	2.28	2.12	2.89	4.65	3.38	3.17	3.18	3.36	3.76	2.56	2.06	3.39	75.51/60.76
Râul Târgului/Râușor	2.03	2.90	1.90	2.09	3.00	4.46	4.02	2.50	2.94	3.53	4.26	1.82	2.10	3.64	50.00/57.69
Cerna/Herculane	5.55	4.22	4.92	5.19	7.66	8.91	11.83	6.23	5.58	7.34	9.95	2.89	4.18	5.79	49.91/72.19
Dunărea/Portile de Fier	5448.62	5463.61	5618.28	3930.36	5451.87	6345.47	6437.68	4696.29	4826.43	5421.63	7602.00	4226.30	4387.11	5533.0	76.38/79.28
Siret/Drăgești	54.53	57.71	98.21	48.86	56.46	104.18	151.59	80.69	133.91	72.60	167.35	49.36	40.76	75.54	65.34/53.95
Prut/Stânca Costești	58.73	84.08	110.27	70.30	74.09	98.90	114.42	72.78	142.44	59.44	183.54	53.77	40.01	84.12	63.92/47.56
Olt/Rm.Vâlcea	101.22	99.55	111.97	84.06	121.99	194.98	148.40	118.99	121.91	121.99	178.50	119.77	80.60	127.60	93.86/63.16

Table 10.5 Analysis of the multiannual average flow rates and of the values considered as hydrological

No	River name	Section	Period of time	Multiannual average flow rate (m^3/s)	Range of the difference considered normally hydrological		Observations
					-10% x multiannual Qmed	$+10\%$ x multiannual Qmed	
1	Lotru	Upstream of Lake Vidra	1950–2012	3.73	3.35	4.10	Of the 63 years analyzed, a number of 24 can be classified as normal
2	Someșul Cald	Upstream Fântânele	1950–2012	7.60	6.84	8.36	Of the 63 years analyzed, 27 can be classified as normal hydrological regime
3	Sebeș	Hydrometic station Petrești	1950–2012	9.19	8.27	10.11	Of the 63 years analyzed, only 10 can be classified as normal to the hydrological regime
4	Bistrița	Upstream Izvorul Muntelui	1950–2012	48.36	43.52	53.19	Of the 63 years analyzed, 18 can be classified as normal
5	Râul Mare	Upstream Gura Apelor	1950–2012	9.11	8.19	10.02	Of the 63 years analyzed, 12 can be classified as normal

(continued)

Table 10.5 (continued)

No	River name	Section	Period of time	Multiannual average flow rate (m^3/s)	Range of the difference considered normally hydrological		Observations
					− 10% x multiannual Qmed	+ 10% x multiannual Qmed	
6	Someșul Mic	Hydrometic station Cluj Napoca	1960–2012	13.93	12.53	15.32	Of the 53 years analyzed, only 14 can be classified as normal. These are: 1960, 1962, 1964, 1965, 1966, 1968, 1989, 1973, 1985, 1997, 2000, 2002, 2008 and 2009
7	Olt	Hydrometic station Cornetu	1967–2012	111.30	100.17	122.43	Of the 46 years analyzed, a number of 9 years can be classified as normal, namely: 1969, 1973, 1974, 1976, 1977, 1985, 1995, 1999 and 2007
8	Argeș	Upstream Vidraru	1950–2014	7.50	6.75	8.25	Out of the 65 years analyzed, a number of 30 years can be classified as normal to the hydrological regime
9	Dunăre	Bazias	1950–2013	5561.64	5005.47	6117.80	Out of the 63 years analyzed, a number of 31 years can be classified as normal

(continued)

Table 10.5 (continued)

No	River name	Section	Period of time	Multiannual average flow rate (m³/s)	Range of the difference considered normally hydrological		Observations
					− 10% x multiannual Qmed	+ 10% x multiannual Qmed	
10	Prut	Rădăuți-Prut	1978–2013	86.84	78.15	95.52	Out of the 35 years analyzed, a number of 3 years can be classified as normal

Note Larger hydrometer stations result in smaller variability in flow rates. Water flows are measured at a higher average altitude (mountainous area) or on a very large catchment area (Danube), so average annual flow rates are recorded in lower variation differences
Data source Chilikova-Lubomirova (2015b)

4. To determine the conditions of duration, intensity, manifestation, effects in order to consider drought as an excessive and natural disaster.

From the conclusion of the expert appointed to this goal, we select the following defining passage:

"Hydro-energy developments must take into account the full range of hydrological risk phenomena: Droughts, floods, etc. When they do not provide funding for accurately tracking them, risk cannot be predicted. In the present case the risk was obvious. The water drought has even been on the Danube, the most important producer of hydropower and the sensor of the impulses it receives from the outside (tributaries). The hydrological analysis during the period 1921–2014 (94 years), for the period September to March (7 months), highlights the occurrence of extreme droughts between 1921–1922, 1946–1947, 1953–1954 (historical), 1990–1991, 2011–2012 (second place)".

In my opinion, this statement is correct, but it has proved that Hidroelectrica Corporation did not take the right steps to manage a time period with implicitly reduced rainfall with a reduced leakage from normal. This could be done by a number of methods and means, with the technical and scientific possibility of mitigating the negative effects of a period of drought.

Let us not forget that an important role of hydropower accumulations is to attenuate major water variations, accumulating water in the intervals with rich run-off and redistributing from the volumes accumulated in the intervals with minimum run-off and requirements of maximum uses.

Minimum liquid leakage phases

The minimum drain is characterized as one of the extremes of the hydrological regime. At the same time, it can be considered as a result or indicator of drought.

In practice, it is of particular interest in areas where there are no accumulation lakes or where the influence of leakage is seasonal or depends on human activity, and the flow regime is widely distributed. In this case, its essential understanding and determination is extremely important. McMahon and Arenas (1982) stated that the minimum leakage is defined seasonally and is directly linked to the annual solar cycle or even to its local or regional climatic effects. The minimum leakage can also be absolute or relative. Different geographic areas are characterized by different behaviors of minimal leakage. For the lower Danube basin with moderate continental climate conditions, the spill is most often due to rain-snow water, and the appearance of two dry seasons is representative. These seasons are summer and winter, the first season being more severe in terms of drought than the second. The seasonality and the severity of the minimum leakage depend on the climatic and physical geography of the basin. It may vary at different time scales, with cases of spicy summers per river section. Taking into account the definition of drought as a period of abnormally dry weather, different parameters of minimum leakage can be observed.

The drought occurs for certain periods of time. There are six types of water-drought degrees, as Beran and Rodier (1985) said:

1. Leakage of between three weeks and three months, with drought during the regeneration and growing of plants, catastrophic for agriculture dependent on irrigation directly from rivers without accumulation lakes.
2. A minimum rate of significant or longer than the normal minimum required during the growing season of the plants. The germination period is not affected by this type of drought and has low consequences for agriculture.
3. A significant deficit of total annual leakage. This affects hydro production and irrigation in large reservoirs.
4. An annual maximum water level below the normal river level. This may introduce the need for pumps for irrigation. This type of drought is linked to the annual leakage deficit.
5. The drought that lasts for several consecutive years, like the "secas" in Northern Brazil. The flow rates remain below the minimum threshold or the rivers completely dry and remain dry for a long period (https://monitordesecas.ana. gov.br/mapa?mes=12&ano=2021).
6. Significant natural depletion of aquifers. This is difficult to estimate, as the identification of the actual level of aquifers is affected by the intensive use of groundwater during drought.

According to McMahon and Diaz Arenas (1982), the following concepts have to be taken into account. The minimum leakage period is usually defined as follows:

• Its duration is often the same as that of the dry season. This occurs when the season is characterized by complete the lack or significantly reduced rainfall; if the absolute or lowest minimum flow rate is almost always equal to the daily minimum flow rate during one year.
• This series of minimum debits expresses the correspondence between fixed durations (expressed in number of days) and flows that have not been exceeded during

a number of consecutive days or not. For example, water flow rates not exceeded for 7–10 days; flow rates not exceeded for 15 days and flow rates not exceeded for one month.

The following cases may affect the process of determining the minimum flow rate—if the water is frozen or if the reserves feeding the riverbeds have been completed or are insufficient to generate surface leakage (although the underground flow continues). Thus, understanding and evaluating both processes and their characteristics are important. Best estimate subjects to volatility adjustment—total (life other than health insurance, including non-life). This is of great importance in terms of the quantitative and qualitative aspects of water resources which are more sensitive during minimum leakage or droughts and in maintaining the ecological health of conditions in aquatic ecosystems, in order to ensure a minimum acceptable flow rate in rivers. For example, there is a strong link between water quantity and quality, and in the case of minimal leakage, the concentration of substances is higher than in the case of average leakage or flash floods. Thus, actual measurements and monitoring must be followed to achieve the actual results. A variety of technical measures may be used, possibly different from those used for average and maximum leakage measurements, but the most important aspect is their choice according to the specific nature of the area under investigation and the hydrological conditions.

Different methods could be used to practically implement monitoring and direct measurement of minimum leakage. Most of them are presented in the review of the World Meteorology Organization (WMO-No. 1044), where the organization of the network for minimum leakage rate studies for water resource development and adequate monitoring of droughts and minimum leakage is well presented. Some of the methods are reflected by Chilikova Lubomirova (2009), Pisota et al. (2010), ISO 438:2008, ISO 4359:2013, ISO 4360:2008, ISO 4374:1990, ISO 4377:2012, ISO 13550:2002, etc. Finally, the data quality analysis follows the procedure for expressing uncertainty and confidence in the measurements taking into account the conditions of the selected case. The procedure is well presented in JCGM 100:2008; GUM (1995); M3003 (2007); ISO/TR 5168:1998, etc.

Minimum water leakage indices

Basic indices of the minimum leakage:

The minimum leakage can be analyzed in different ways of analyzing the time series of daily flows in order to produce summary information describing the minimum leakage regime of a river. The term "minimum leakage indices" is used for specific values derived from a minimum flow analysis. Some are unique values such as the recession constant, the base rate index or the average of a time series. These are called minimum leakage rates. More complex methods estimate the probability of minimal leakage. For example, daily flows' cumulative frequency distribution (flow duration curve) describes the relationship between flow and time in percent when a given flow rate is exceeded. The theory of extreme values is used to estimate the probability of annual minima not being exceeded. What is essential in the two techniques is that the leakage duration curve takes into account all days in a chronological series and thus the duration in percentage of the entire observation period in which a flow rate is exceeded. In contrast, the theory of extreme values applied to the annual minimum series estimates the probability of non-exceedance in years or the average interval in years (recovery period) when the annual minima are below a given value. Therefore, it is often helpful to specify the water resources management plan (WMO-No. 1029).

Many water resource management decisions are based on these indicators:

- Average water flow rate (m^3/s)—is one of the most commonly used hydrologics and water resource planning statistics. It can be estimated from a chronological series of measurement data by summing all daily flows and dividing by the number of days with observations. It is normally calculated for a calendar year or for the hydrological year of the time series. It may also be calculated for specific months or seasons.

The situation at Romania level, especially in the water catchment areas where hydropower facilities are operated by Hidroelectrica Corporation are found, is analyzed in terms of minimum average flow rates (Table 10.6).

- 95% assurance rate (Q95%) (cubic meters/s)—is one of the most commonly used minimum leakage indicators in an operational way and is defined as the rate that is exceeded in 95% of the time interval. It can be determined by ranking all debits (daily, monthly, yearly) and finding the flow rate with 95% of all values of the analyzed data line. This percentile and also others (Q90%, Q70%, etc.) can be determined from the leak-proof curve (Table 10.7).

Statistical methods and approaches for assessing minimum water leakage

The minimum leakage regime of a river can be analyzed in a variety of ways, depending on the initially valid data and the type of output required information (Chilikova-Lubomirova and Dimitrov 2012). Therefore, there is a wide range of indices and minimum flow measurements. The term "minimum leakage rate" used here refers to the different methods that have been developed to analyze, often in graphic form, the minimum leakage rate of a river. The term "minimum leakage rate" is used most often to define particular values obtained from any minimum leakage measurement (sometimes, it is difficult to separate them from each other) (Smakhtin 2001).

The statistical analysis of the minimum leakage indicates the availability of water in whites at times when the water requirements for use are likely to be out of reach. For this reason, minimum leakage methodologies are required by all fora in charge of water resource management for water planning, management and regulatory activities. These activities include:

- the development of environmentally effective river basin management plans;
- the support for and permit new water extraction, interbasal transfers and wastewater flows;

Table 10.6 Minimum average characteristic flow rates recorded in the range 1950–2012

No.	Hydrometer station	River name	Flow rate multianual (m^3/s)	Flow rate minimum recorded in range1950–2014 (m^3/s)/year
1	Baziaș	Danube	5561	3773/1990, 3932/2003, 4167/1950, 4213/2011, 4285/1993, 4373/1971, 4397/2012, 4432/1961, 4480/1973, 4578/1983
2	Lotru	Vidra Lake	3.72	1.67/2011, 1.91/2012 2.36/1973, 2.42/1993, 2.43/1990, 2.68/1997, 2.80/1963, 2.82/1950, 2.83/2000
3	Râul Mare	Gura Apei	9.11	2.87/2011, 3.05/2012, 4.83/1990, 5.81/1993, 6.15/1985, 6.53/1950, 6.80/1954
4	Radăuți-Prut	Prut	86.84	34.15/1990, 40.63/2012, 42.07/1987, 48.57/1994, 56.74/2011
5	Fântânele	Someșul Cald	7.31	2.31/1992, 4.06/1961, 4.37/1954, 4.71/1950, 8.25/2012, 10.31/2011[a],
6	Vidraru Lake	Argeș	7.51	4.71/1997, 4.87/1963, 6.84/2012, 9.19/2011[a]
7	Fântânele	Someșul Cald	7.60	2.31/1992, 4.06/1961, 4.37/1954, 4.71/1950, 5.37/1983, 10.29/2011[a], 8.25/2012[a]
8	Straja	Bistrița	7.47	1.99/1986, 2.55/1990, 3.12/1994, 3.52/2012, 3.63/1987, 5.21/2011
9	Izvorul Muntelui Lake	Bistrița	48.36	24.90/1950, 25.26/2012, 28.80/1963, 29.09/2011, 29.4/1987, 30.0/1990, 30.6/1954
10	Cluj Napoca	Someșul Mic	13.93	4.66/1992, 6.19/1961, 8.37/1994, 8.50/2012, 9.43/2003, 11.35/2011

[a]The average annual flow rates recorded were above the section-specific multiannual average value.
Data source National Institute of Hydrology and Water Management

Table 10.7 Typical average minimum rates of credit with different degrees of insurance

No.	Hydrometer station	River name	Flow average multiannual (m³/s)	Flow average 2011 (m³/s)	Flow average 2012 (m³/s)	Degree of insurance—exceeding of liquid leakage		
						Flow average 80% (m³/s)	Flow average 90% (m³/s)	Flow average 95% (m³/s)
1	Baziaș	Dunăre	5561	4213	4397	4738	4408	4132
2	Lotru	Upstream Vidra	3.72	1.67	1.91	3.20	2.96	2.77
3	Râul Mare	Gura Apei	9.11	2.87	3.05	6.57	5.36	4.40
4	Radăuți-Prut	Prut	86.84	56.74	40.63	55.30	44.41	37.71
5	Fântânele	Someşul Cald	7.31	10.29	8.25	5.98	5.20	4.59
6	Vidraru Lake	Argeş	7.51	9.19	6.84	–	–	–
7	Straja	Bistriţa	7.47	5.21	3.52	3.74	3.04	2.65
8	Petreşti	Sebeş	9.19	8.41	5.41	6.31	5.21	4.57
9	Izvorul Muntelui Lake	Bistriţa	48.36	29.09	25.26	37.35	32.36	28.77
10	Cornetu	Olt	111.3	99.06	69.88	81.79	68.71	58.38
11	Cluj Napoca	Someşul Mic	13.93	11.35	8.50	10.26	8.68	7.50

Data source National Institute of Hydrology and Water Management, Romania

- the determination of threshold values for minimum flow rates for the maintenance of the aquatic biosphere;
- the land use planning and regulation.

Commercial, industrial and hydroelectric entities also require minimum water leakage methodologies to determine the availability of water for water supply, wastewater flow and energy production (Kernell and Friesz 2000).

As regards international studies on the minimum leakage and drought, Tallaksen and van Lanen (2004) claim that there is extensive literature on the various processes that operate during the minimum leakage or a period of drought. In particular, the basin response and the minimum leakage regime of a river or region have been described. Less material is available about assessing minimum leakage and drought methods, including prediction, forecast and estimation in unmonitored sections. The latter has already been stipulated in research relevant to the forecast of the minimum leakage for the hydrometer-controlled sections and is still valid (Chilikova-Lubomirova 2015b; Tallaksen and van Lanen 2004).

Smakhtin (2001) concluded that in the context of an integrated and environmentally sustainable management of the river basin, the minimum flows could be seen as a dynamic concept rather than described by only one characteristic of the minimum leakage. Therefore, priority are the flow rate time series from which the variety of minimum leakage indices can be extracted to meet different management and engineering purposes.

The most common regionalization analysis of leakage is the methods for the duration of the leak, including, inter alia, the base leakage index, the leak rate analysis and the recessionary parameters. The "Manual on minimum leakage estimation and forecasting", published by the World Meteorological Organization, provides a comprehensive summary of how to analyze the series of liquid flows, especially the minimum flow rates. Although the indices and methods for calculating the minimum leakage have been well documented in this manual, a comprehensive software is missing which provides a quick and standardized calculation of the minimum leakage statistical analysis. This software is based on the open-source R statistical program and extends to the analysis of the daily average flow time series, allowing a fast and standardized calculation of the minimum leakage analysis. There are different analyses of the time series of daily average flows to produce concise information describing the minimum drainage regime of a river. This application package provides functions to calculate the described statistics and produces graphs similar to those in the manual. The authors of this software are Koffler and Gregor (2013). In Romania, researchers from the National Institute of Hydrology and Water Management (NIHWM) collaborated with the European FRIEND Working Group—minimum water leakage and drought on the topic "minimum leakage rates" to prepare a large publication about the current minimum leakage conditions in Europe. The study consisted the calculation of the leakage of minimum indices based on the above-mentioned package of applications.

On the basis of the results obtained, as a result of the use of this package of applications, indices of minimal leakage have also been calculated. These statistical methods and approaches can certainly be a comprehensive and essential analysis in assessing the minimum leakage.

5. Establish the moment in which the event invoked by Hidroelectrica Corporation as being a force majeure event could have been reasonably anticipated by a specialist in the field taking into account the statistics referring to the previous periods of time and the available estimations regarding the following periods.

The conclusion of the expert appointed to that objective shall be as follows: "The time can be detected up to 1 month before ($\pm$ 10% deviation). For larger periods the deviations are $\pm$ 40–50%. The inflow forecast for the first of the three months is the basis of the operational program (spent flows) and determines with a high degree of precision ($\pm$ 10%) the output for the following month. The forecast of the last two months is indicative. For this reason, the values of predictable energy over the last two months of the forecast are used only to compare with already contracted values and make decisions on future contracting policy". In my opinion, the expert-designate has only partially responded to the objective. The reasons for this opinion consist of the following arguments.

At the written request of the economic operators to the Romanian Chamber of Commerce and Industry (RCCI), the Law and Law Department, the Legal Office, the existence of force majeure is hereby approved on the basis of Article 28(2)(I) of Law 335/2007. The Romanian Chamber of Commerce and Industry issued opinion No. 1240 on 27.09.2011 at Hidroelectrica's request that force majeure be invoked, but an act not signed by the president of the National Chamber was declared void in Resolution No. 5 of 07.11.2011. This resolution in turn was subsequently declared void by the RCCI.

Where an economic operator invokes force majeure, the request shall include a factorial and detailed account of the event, its consequences in relation to the contracting partner and legal arguments that the alleged event is force majeure.

An opinion on the existence of force majeure shall be issued on the basis of supporting documents submitted exclusively by the applicant. The documents shall contain at least: the trade contract affected in the event of force majeure, including the force majeure clause; certificates from the competent bodies, authorities and institutions, on a case-by-case basis (other than the Romanian Chamber of Commerce and Industry), on the existence and effects of the alleged event, its location, the time when the event started and ended; notifications to the counterparty regarding the occurrence of the alleged event and its effects on the conduct of the contractual operations.

The documents shall be submitted in the original or in a copy certified by the applicant. The applicant shall assume full responsibility, including criminal liability, for the veracity of the data and documents submitted with the request to the Romanian Chamber of Commerce and Industry for the endorsement of the force majeure event.

Given that the contract between the two parties does not exist in the chapter on force majeure, references to the value of the water flows/level of the lakes/and more that can be analyzed in the hydrological context of the years 2011 and 2012 and multiannual by a specialist in the hydrology cannot be given a view of this objective.

The documents located at the Romanian Chamber of Commerce and Industry justifying the case of force majeure in which Hidroelectrica Corporation was involved include a material presenting the situation of the hydrological regime on the Danube and inland rivers, a material issued by the Emergency Inspectorate and one by the Romanian Waters. In these documents, general data are presented at country level without any particularities on river basins depending on the production share of hydropower units.

6. Indicate whether or not Hidroelectrica Corporation is unable to fulfill its contractual obligation due to the event being invoked as force majeure, or whether because it was too difficult to do so and too onerous in order to avoid damage.

The conclusion of the expert appointed to this objective is that: "the situation of the generalized hydrological drought on the entire territory of Romania, including on the Danube River, makes us think that it could be a case of force majeure, which entails an impossibility to fulfill the contractual obligations to produce electricity. Meteorological and hydrological droughts represent risk phenomena and have a negative impact on any economic activity based on water use. The phenomenon of extreme drought could not be taken into account on the basis of long-term forecasts (see when the contract was concluded). The average multiannual and annual flows figures recorded in 2011 and 2012, or multiannual flows figures for the months concerned, show a clear reduction for the period 2011 to 2012". In my opinion, the expert-designate did not make a correct (real) response to the objective.

The reasons for this opinion consist of the following arguments. The situation of the hydrological drought phenomenon, which has occurred on the territory of Romania, including the Danube River basin, especially in some of the time periods mentioned above, makes us believe that it could not be a case of force majeure, which entails an impossibility of fulfilling contractual obligations to generate electricity, because the very rich spill in the previous year 2010 and the spill over the multiannual monthly averages of months in 2011 and 2012 could generate water supplies to be used in poor times. Meteorological and hydrological droughts are risk events and have a negative impact on any economic activity based on direct water use (in the case of microhydropower plants), which is not the case for hydropower accumulations administered by Hidroelectrica Corporation, which are due to the useful volumes of the reservoirs currently available, it can also redistribute significant volumes of water over time (water accumulation in rainy periods and redistribution into dry periods, flood wave attenuation).

In support of Hidroelectica's claim that it claims force majeure, it also argues that in early 2011, the level of filling of lakes was very low, such as Oasa—52%, Fântânele—58%, Dragan—22% or Vidra—43%. Hidroelectrica Corporation's production capacity was well above the contractual obligations with its partners at the time (Energy Holding trading company, Alpiq RomIndustrie and RomEnergie trading company, ALRO Corporation, Electrocarbon Corporation et al.).

The difficulty in fulfilling contractual obligations is, in our view, caused by the combination of drought in some river basins (which was not exceptional or unforeseen). Poor management of hydropower accumulation led to a drastic decrease in the amount of energy produced by Hidroelectrica Corporation.

10.2   Questions Put Forward by Hidroelectrica Corporation

1. **The hydrological definition of the term "normal hydrological year".**

The expert appointed to this objective concludes that: "The normal hydrological year is the year in which the average annual flow rate is close to the multiannual average of this variable (± 10%). In order to be representative, the average annual water flow shall be calculated over a long period of time, generally at least 30 years of observations. The forecast has high assurance if the expected time period is long. If the data string is reduced the error rate is high. The forecast flow rate may be higher or lower than 1: Over-unit value if a certain time interval is richer (> 1); sub-unit value if the time interval is poorer (< 1)".

In my opinion, this conclusion can be complemented by the following points: For a few inland rivers and for the Danube when it enters the country, the values of the flows considered in "**normal hydrological year**" range have been calculated, with summary data presented previously (Table 10.3).

2. **Hydrological characterization of the period between 30.09.2011 and 30.04.2012.**

From the conclusion of the expert appointed to this objective I have taken note of the following ideas: "The average flow rate in 2011 and 2012 was lower than the average flow rate of the annual flow rate for all rivers equipped for the production of electricity. The lowest values have been found on the Lotru, Sebes, Cerna, Bistra, Târgului and Prut rivers where values reach 50% of the average annual flow. The other rivers account for 60–70% of the average annual flow rate. The values for both years are relatively close. The hydro plants located on the Danube river, the largest producer of hydropower, is worth 76.38% in 2011 and 79.28% in 2012".

...

Drought does not directly affect the production of electricity during its production. It can have future repercussions if it is extended for a long time. If the previous year (or years) had rich flows, electricity can also be produced in dry periods when water has been preserved (only if the requirements have been reduced).

Between October 2011 and April 2012, Hidroelectrica was unable to deliver electricity to its full capacity as the inflow rates were much lower than the average multiannual rates and lower than the forecast.

In my opinion, the expert-designate did not make a clear conclusion to this objective, showing whether Hidroelectrica Corporation other options to exploit its capacities, so I feel obliged to inform you of the following:

The authors analyzed the range 30.09.2011–30.04.2012 in the sections: upstream accumulation of Vidraru–Arges River, Cluj Napoca–Somesul Mic, Cornetu–Olt River, Petresti–Sebes River; upstream accumulation of Fantanele–Somes River warm; upstream accumulation of Mount River–Bistrita River; upstream accumulation via water–Raul Mare River; upstream accumulation Vidra–Lotru River, hydrometic station Straja–Bistrita River, hydrometric station Radauti-Prut–Prut River; and hydrometic station Bazias–Danube River, on the basis of monthly average flows rates from 1950 to 2012 (a much longer period than that of the expert appointed by the court using water data from the range 2000 to 2012).

Liquid leakage (water flows) from the time period 30.09.2011 to 30.04.2012 was compared with the corresponding multiannual monthly average water flow to establish the situation of deviations from a situation considered as "normal", but also with the monthly average water flow in the years characterized as deficit (drought) to see if the phenomenon recorded in 2011–2012 was observed; it was an unpredictable one of extreme magnitude (Fig. 10.3).

The monthly average flows were below the multiannual monthly average in a higher or lower proportion (Fig. 10.3). Looking at the variation of their monthly average in 2011 compared to the synthetic hydrograph of the leak at the hydrometric station Bazias, it can be observed that in January, the value of the leak is much higher than the monthly multiannual average. A volume of water could be accumulated in this month for later use in the dry months of the hot season. February was a normal month with similar amounts of leakage in 2011 (Fig. 10.4). Instead of an increase in average water flow values, they fall until August when the 2011 leakage value is comparable to the multiannual average, leading to a downward trend thereafter.

The year 2012 (Fig. 10.5) is characterized only by a smaller leakage than normal, but not by much compared to other years. This reduced leakage is a hydrological phenomenon occurring on any river (in Romania or within this temperate-continental climate). From Fig. 10.6, it is possible to identify the periodicity of the low drain intervals that have taken place on the Danube in the analyzed range from 1950 to 2013. This shows deficit years from 10 to 10 years, such as 1950, 1961, 1971 and

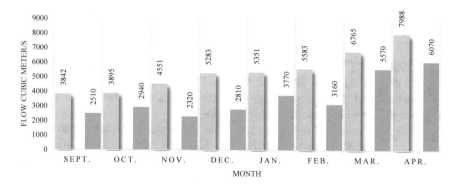

Fig. 10.3 Comparative chart monthly average flows range 30.09.2011–30.04.2012 and average monthly rates recorded at the hydrometric station Bazias–Danube River

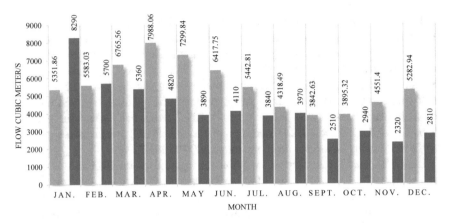

Fig. 10.4 Average monthly flowchart of 2011 and average monthly flow rates recorded at the Bazias–Danube River hydrometric station (Series 1—flow rates year 2011; Series 2—average monthly flow rates multiannual 1950–2012)

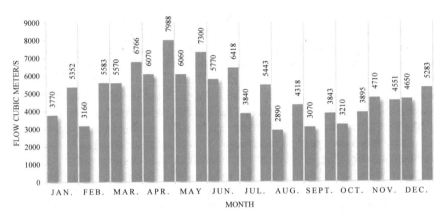

Fig. 10.5 Comparative chart monthly average flows from 2012 and monthly average rates recorded at the Bazias–Danube River hydrometric station (Series 1—flow year 2012; Series 2—monthly average monthly rates 1950–2012)

1973, 1983, 1990, 1993, 2003, 2011, where the least leakage occurred in 1990 is 3773.33 m^3/s compared to 4213.22 m^3/s in 2011 or 4397.5 m^3/s in 2012.

From our point of view, the phenomenon has not been unpredictable or exceptional in nature that can be invoked as a force majeure.

In the case of the Arges River (analyzed in the upstream section Vidraru accumulation), it is noted that in the period 30.09.2011–30.04.2012 (Fig. 10.7), the differences between the monthly average recorded flows and the multiannual monthly average flows are not significant. The time intervals when the average monthly flow rates have slightly or significantly exceeded the average multiannual value have resulted in a bigger leak than the monthly usual.

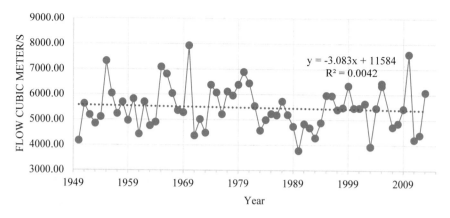

Fig. 10.6 Change in the average annual flow rates of the Bazias hydrometric station in the range from 1950 to 2013

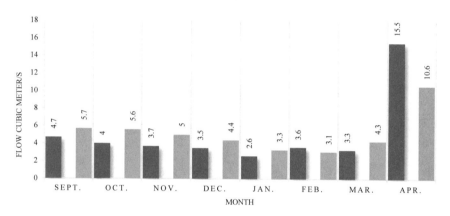

Fig. 10.7 Average monthly flowchart interval 30.09.2011–30.04.2012 and average monthly flow rates recorded in upstream of Vidraru–Arges River accumulation (Series 1—annual flow rates 2011–2012; Series 2—monthly average monthly flow rates 1950–2014)

The analysis of Figs. 10.8 and 10.9 also shows that the failure to achieve the amount of program energy was also due to poor management of the build-up over a period of at least 3 months (for which the hydrological forecasts from NIHWM were available), where no accumulation of volumes of water in lakes was made for use in the deficit months, although the respective level/volume of water retained in the lakes at that date allowed this (Fig. 10.10).

In the case of Vidraru accumulation, the average annual flows recorded in 2011 and 2012 do not reflect the need to invoke force majeure and to issue a document by the Romanian Chamber of Commerce and Industry, including the Arges River basin (Fig. 10.11).

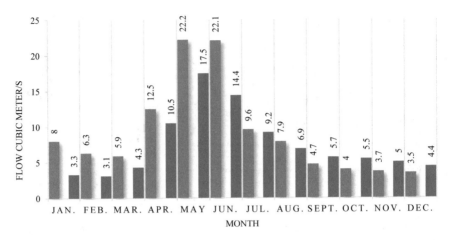

Fig. 10.8 Comparative chart of monthly average debits for 2011 and multiannual average monthly debits recorded at hydrometric station upstream Vidraru Lake–Arges River (Series 1—year 2011 flows; Series 2—monthly multiannual average flows 1950–2014)

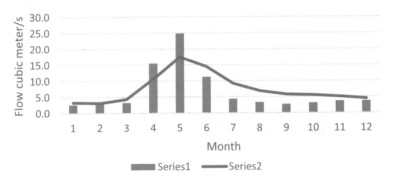

Fig. 10.9 Comparative chart of monthly average flows from 2012 and multiannual average monthly debits recorded at hydrometric station upstream Vidraru Lake–Arges River (Series 1—year 2011 flows; Series 2—monthly multiannual average flows 1950–2014)

On the River Prut, there is a decrease in the monthly average flow rates (Fig. 10.12), but with a maintenance of the variation similar to the multiannual average variation regime. However, the annual average value recorded at this hydrometric station (HS) is not an exceptional one, the lowest flow rate having been recorded in 1990 at 34.15 m^3/s (Figs. 10.13 and 10.14).

The forecast of these low drain periods as well as of a year characterized as a drought had to be made especially since the historical experience in the Prut River basin never had two years in succession with a rich liquid leak. So, after 2010 (year with a very rich drain and record energy production for Hidroelectrica Corporation), one year could be recorded with much lower values, or some below the multiannual average of the leakage in the section under review.

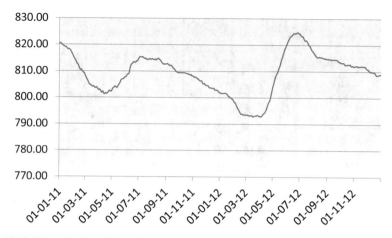

Fig. 10.10 Water level evolution in the hydropower accumulation Vidraru–Arges River. *Data source* Hidroelectrica Corporation

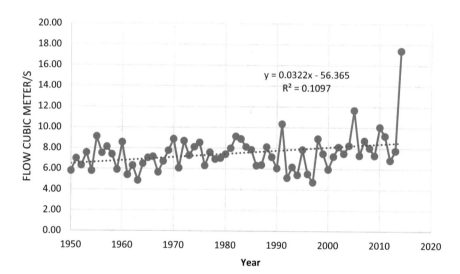

Fig. 10.11 Variation of the average annual flow rates at Vidraru Lake–Arges River in the range from 1950 to 2014

The same situation as that of the River Prut is found on the River Lotru, where the liquid leakage was analyzed in the section upstream of Lake Vidra (Figs. 10.15, 10.16 and 10.17).

In the case of this river, the lowest flow rates are among the lowest overall. With the data provided by the operation service of Hidroelectrica on the share of project energy of the mean hydrological year, the Vidra accumulation on the Lotru River does

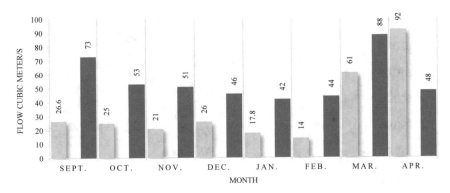

Fig. 10.12 Comparison chart monthly average flows range 30.09.2011–30.04.2012 and monthly average monthly rates recorded at Raduti-Prut River HS (Series 1—beginning year 2011–2012; Series 2—monthly average monthly rates 1978–2013)

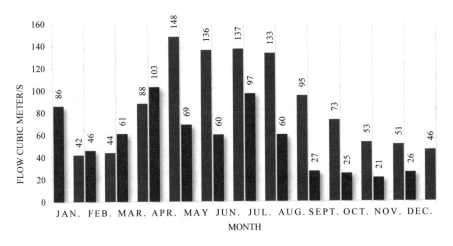

Fig. 10.13 Comparative chart monthly average flows from 2011 and monthly average monthly flows recorded at Radauti–Prut River HS (Series 1—beginning year 2011; Series 2—monthly average flows 1978–2013)

not have a great influence on the total amount of energy produced by Hidroelectica SA.

3. **The expert needs to mention if, taking into account the hydrological characterization of the presented interval of time, Hidroelectrica Corporation had other options of exploiting the production capacity outside of the one already followed, thus not jeopardizing the safe functioning of the National Energetic System while also causing minimal losses to contractual partners.**

The expert appointed to this objective concludes that "Although 2010 was rainy, Hidroelectrica could not make important water reserves because not all production

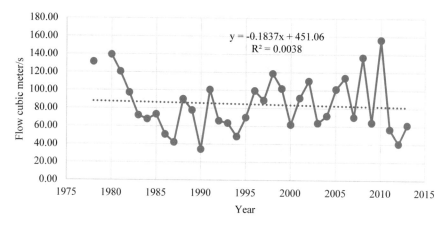

Fig. 10.14 Variation of the annual average flow rates at Radauti-Prut HS–Prut River in the range of 1978–2013

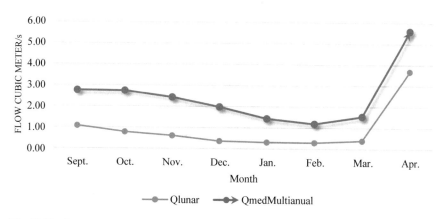

Fig. 10.15 Comparative chart average monthly flows range 30.09.2011–30.04.2012 and average monthly multiannual debits recorded at upstream Vidra Lake HS–Lotru River (Series 1—annual flows 2011–2012; Series 2—monthly multiannual flows 1950–2012)

capacities had estimations for such events. Additional water build-up in the dam lakes can only be made on the River Bistrita at hydroelectric power station Stejaru, on the Arges River at hydroelectric power station Corbeni, on the Lotru River at hydroelectric power station Ciunget, on the Somes River and hydroelectric power station Mariselu and on the Sebeș River and hydroelectric power station Gâlceag. No extra accumulation can be made on the Danube due to high flow rates and exceptionally large volumes covering large land areas. The production of a normal amount of

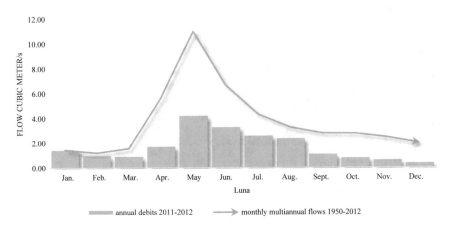

Fig. 10.16 Comparative chart of monthly average flows for 2011 and multiannual average monthly debits recorded at upstream Vidra Lake HS–Lotru River (Series 1—annual debits 2011–2012; Series 2—monthly multiannual flows 1950–2012)

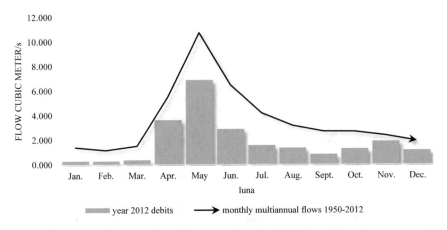

Fig. 10.17 Comparative chart of monthly average flows for 2011and multiannual average monthly flows recorded at upstream Vidra Lake HS–Lotru River (Series 1—year 2012 debits; Series 2—monthly multiannual average flows 1950–2012)

energy at the beginning of the drought period may be due to the additional accumulation of water in the above facilities. Emptying them during the period of drought has resulted in a decrease in water volume and thus in lower electricity production".

Considering the fact that before the interval subject to the analysis 30.09.2011–30.04.2012, in Romania at the level of its river basins, was registered in the decade 2000–2010 the high average value of the flows at the level of all rivers (in the years 2005, 2006, 2008 and 2010, there were historical floods in Romania) and the fact that the company Hidroelectrica Corporation has in its portfolio water accumulations with high retention volumes, which can be used as water storage tanks in the intervals with a rich drain that could later be machined (redistributed) during dry periods with low leakage.

4. **The expert has to identify the hydrological forecasts available to Hidroelectrica Corporation and specify: (i) the degree of accuracy for periods of 1 year and (ii) the maximum period of time for which such forecasts can be considered as having a high degree of certainty.**

The conclusion of the expert appointed by the court to this objective is: "water forecasts are often addressed to short-term events (floods) or longer-term events (seasonal). The very long ones, for example 12 months, have a high degree of error (30–40%). The accuracy and timeliness of hydrological forecasts shall depend on the reliability and quantity of hydrological and meteorological information, the response time of the river basin, the rate at which the status of the basin can be assessed at any given time, the forecasting techniques or models used, The speed with which the beneficiaries receive the forecast, etc. (Diaconu and Jude 2009). Regardless of the type of prognosticated element, the emitted provisions imply a grade of error expressed in percentages of the true value or absolute value (m^3/s, m^3, days, months, etc.).

Long-term (1 year) average flow forecasts have an error rate of 40–50%".

In my opinion, the expert-designate has only partially responded to the objective. The reasons for this opinion shall consist of the following arguments put forward in the first objective by the appellant.

My impartial conclusion after a detailed analysis of the phenomena specific to the objectives is that:

Difficulties in meeting contractual obligations are due to the combination of drought in some river basins (which was not exceptional or unforeseen), with poor management of hydropower accumulation leading to a sharp decrease in the amount of energy produced by Hidroelectrica Corporation and a contracting of production to a greater extent than could be delivered under conditions of natural variation of the hydropower-operated river regime.

Technical Adviser chosen by the impugners
DDC

10.3 Post Factum

The reference to the force majeure clause for the purpose of suspending/cancelling contracts for the supply of electricity to different contract partners by Hidroelectrica Corporation was made on the basis of variations in the hydraulic nature of the hydroenergy threatened sections, from surplus to deficit.

In denouncing the contracts, no motives were invoked that would have put the society in the capacity to produce and supply the electric energy to the beneficiaries.

Regarding the situation after the technical experts appointed by the parties and the expert appointed by the Bucharest Court, there is a lack of clarity and vision in drawing up the electricity supply contract regarding the Article of force majeure.

The provisions of this article should have referred mainly to extreme characteristics of the hydrological regime, which could negatively impact energy production.

The hydrological studies carried out in the hydropower sections from 1930 onward and the experience of using day-to-day accumulation should have provided specialists in society with the necessary data to determine the sensitive values at which production falls below a certain limit. The termination of contracts and disputes in national and international courts have also resulted in a lack of a multi-disciplinary approach to the factors involved in electricity production, such as hydrological factors, changes in the storage capacity of water volumes, technical parameters of energy production facilities, optimization and automation of production systems, matching and matching production to market requirements and degree of training of service and management personnel.

References

Adler M-J (1994) Low flows methodologies and characteristics in Romania. Rom J Hydrol Water Resour 1(2):1994

Adler M-J, Chelcea S (2014) Climate change and its impact in water resources in Romania. In: Proceedings of XXVI conference of the Danubian countries on hydrological forecasting and hydrological bases of water management, 22–24 Sep 2014, Deggendorf, Germany

Beran MA, Rodier JA (1985) Hydrological aspects of drought. In: UNESCO–WMO studies and reports in hydrology, vol 39, p 149

Chilikova-Lubomirova M (2009) Meteorological evaluation of methods for discharge measurements and observation. PhD thesis, Sofia

Chilikova-Lubomirova M (2010) Hydrological data. Hidroelectrica Corporation, Bucharest

Chilikova-Lubomirova M (2014a) Hydrological data. Hidroelectrica S.A., Bucharest

Chilikova-Lubomirova M (2014b) Challenges in the hydrologycal indices implementation in the process of drought identification and monitoring. In: 14th GeoConference on water resources, forest, marine and ocean ecosystems. International multidisciplinary scientific conferences SGEM, conference proceedings, pp 863–871, 17–26 July, Albena, Bulgaria

Chilikova-Lubomirova M (2015a) Hydrological and energy data. Hidroelectrica Corporation, Bucharest

Chilikova-Lubomirova M (2015b). Hydrological data. National Institute of Hydrology and Water Management

Chilikova-Lubomirova M, Dimitrov D (2012) Automatic system for drought identification in river basins in Republic of Bulgaria with the SRI index. Bul J Meteorol Hydrol 5(17)

Diaconu C, Șerban P (1994) Sinteze și regionalizări hidrologice. Editura Tehnică, Bucharest

Diaconu CD, Jude O (2009) Hydrological forecasts. Publishing house Matrix Rom, Bucharest

GUM (1995)

https://monitordesecas.ana.gov.br/mapa?mes=12&ano=2021

ISO 13550:2002

ISO 4359: 2013

ISO 4360: 2008

ISO 4374:1990

ISO 4377:2012

ISO 438:2008

ISO/TR 5168:1998

JCGM 100:2008

Kernell GR III, Friesz PJ (2000) Methods for estimating low-flow statistics for Massachusetts streams. Water-resources investigations report 00-4135

Koffler D, Gregor L (2013) LFSTAT—Low-flow analysis in R. EGU general assembly 2013, held 7–12 April, 2013 in Vienna, Austria, id. EGU2013-7770

Laaha G (2002) Modelling summer and winter droughts as a basic for estimating river low flows. In: van Lanen HAJ, Demuth S (eds) FRIEND 2002—regional hydrology: bridging the gap between research and practice. Proceedings of the fourth international conference on Cape Town, South Africa, pp 289–295. IAHS Publ. 274 IAHS Press, Wallingford, UK

Li K, Makarau M (eds) (1994) Drought and desertification: reports to the eleventh session of the commission for climatology. WMO/TD 605, Geneva

M3003 (2007)

Mătreață M (1997) Dinamico-statistic model for low flows. Hydrology studies, NIHWM, 24-37

McMahon TA, Diaz Arenas A (1982) Methods of computation of low streamflow. In: A contribution to the international hydrologycal programme, UNESCO, Imprimerie de la Manutention, Mayenne

Pisota I, Zaharia L, Diaconu DC (2010) Hydrology, Universitara Publishing House, Bucharest 978-606-591-146-1

Smakhtin VU (2001) Low flow hydrology: a review. J Hydrol (240):147–186. http://www.ivsl.com

Tallaksen L, van Lanen HAJ (eds) (2004) Hydrological drought. Processes and estimation methods for streamflow and groundwater. Hydrology and quantitative water management WIMEK

WMO-No. 1029; Operational hydrology report (OHR)-No. 50

WMO-No. 1044, published in 2010nt the manual provides information about stream gauging techniques

www.inhga.ro/web/guest/prognoza_medie_durata_rauri

Chapter 11
Electricity, Systemic Crises and Environmental Policies

Abstract Today's society is facing several crises simultaneously (climate, health, economic, political and humanitarian) which is reflected in the evolution of energy prices. The global dependencies of economic, social and political systems have led to a global manifestation of these crises. Thus, the scale and effects produced are difficult to anticipate and manage in this context of globalization. The energy system is also subject to these crises, which in turn generate other shock waves on society. Under such conditions, energy price trends have become extremely volatile, strongly influencing the budgets of household consumers and, above all, the activities of the energy-intensive industrial sectors. Manufacturing companies, governments, continental or global organizations have tried to find ways of giving themselves time to find medium- and long-term solutions.

Keywords Electricity · Energy resource · Crises

The top electricity producers have changed over time. While in the 1990s the, United States produced 3219 TWh, followed by Russia with 1082 TWh and Japan with 871 TWh, in 1994 Japan overtook Russia and was closely followed by China with a production of 928 TWh, and only a year later, the two countries were joined by each other.

The following years only see an increase in the amount of electricity produced, with the United States reaching 4350 TWh, China 3282 TWh and Japan 1142 TWh. In 2011, China overtakes the United States, reaching 4716 TWh and in 2016, 6200 TWh. In 2021, China produces 8537 TWh, which undoubtedly shows the investments made over time in electricity generation sources (https://ec.europa.eu/eurostat/statistics-explained/index.php?title=Electricity_production,_consumption_and_market_overview). Thus, thermal power accounts for the largest share of the produced mix with 5646.3 TWh, followed at a distance by hydropower 1340.1 TWh, wind power 655.6 TWh, nuclear power 407.5 TWh and solar power 327 TWh (https://www.statista.com/statistics/302233/china-power-generation-by-source/).

Global energy security means more than access to energy with minimal environmental impact. World organizations (G-20, G7) and governments are concerned

© The Author(s), under exclusive license to Springer Nature Switzerland AG 2023
D. C. Diaconu, *Force Majeure in the Hydropower Industry*,
https://doi.org/10.1007/978-3-031-27402-2_11

about protecting the global economy against potential threats such as political and economic threats, socio-economic destabilization (Azzuni and Breyer 2018).

After withdrawing from the Paris Agreement on Climate Change and lifting the ban on crude oil exports, the United States has shifted its priorities from energy security to energy dominance (Nukusheva et al. 2020). Russia's energy policy, on the other hand, tends to emphasize technical rather than abstract, cohesive goals (Salonen 2018). While this approach may prevent fundamental structural changes in the Russian energy sector, it hinders the potential of renewable energy production, especially in the Arctic regions. Moreover, the economic and political measures adopted during 2022 at EU level on Russia have exacerbated the energy crisis, with the prospects for recovery much more pessimistic.

Turkey is the fastest-growing energy market among The Organisation for Economic Cooperation and Development (OECD) countries.

At European level, electricity production has been on an upward trend since 1990 (2132 TWh) until the global economic crisis in 2008 (2844 TWh). After this crisis, there was a slight recovery to 2800 TWh before a new downward trend in 2020 (2664 TWh) caused by bottlenecks due to the COVID-19 pandemic (https://ec.europa.eu/eurostat/statistics-explained/index.php?title=Electricity_production,_consumption_and_market_overview).

The energy policy of European countries such as Poland, Czech Republic, Slovakia and Hungary is closely linked to energy imports (mainly from Russia), which leads to slightly different economic and political measures compared to other EU countries. In this case, switching to supply from other sources such as Middle East or North Africa or renewable sources would be a viable option (Dyduch and Skorek 2020; Nyga-Łukaszewska et al. 2020).

This makes the co-dependence between producers, transporters and consumers much more evident. In this context, the aim is to highlight the crises and their effects in the current energy sector. In a modern energy market, participants must build relationships to prevent conflicts and to implement a common environmental and socio-political policy.

In the explanatory dictionaries, the definition of crisis is presented as "a phase in the development of a society marked by great difficulties (economic, political, social, etc.)". Given the diversity of its forms, specific definitions have also been developed, such as: economic crisis—stagnation and disruption of economic life; energy crisis—a complex phenomenon manifested by a shortage of conventional energy resources; monetary crisis—an economic phenomenon characteristic of the capitalist regime, consisting of a lack of money and credit on the market due to the paralysis of economic life in times of business stagnation; political regime crisis—an era in which a regime that has become unstable is about to be overthrown (by revolutionary means) and replaced by another; government (or ministerial) crisis—the period of time (characterized by political turmoil and unrest) between the resignation of one government and the formation of the next; or simply a product crisis—a shortage of goods on the market (Keeler 1993; Boin et al. 2005; Coyne 2011; Grossman 2013).

These crises that a society goes through lead to its shaping, to the reorientation of the functioning of its component systems. According to Congleton (2005), a crisis has three characteristics: surprise, inconvenience and urgency. But because there is "a high degree of uncertainty… both about the nature and potential consequences of the threat", crises can also lead to a sense of helplessness "what next?" and "what can we do?" as well as a tendency toward political recklessness (Boin et al. 2005), hence the generation of force majeure situations.

11.1 Today's Energy Challenges

The globalization of the economy has meant that the shock wave of the energy shortage is being felt on other continents. This has led to energy shortages, tariff increases and general falls in demand which ultimately lead to stagnation in global economic growth (Fig. 11.1).

The disruption of the flow of fossil fuels, which are still largely used to generate electricity, has led to high volatility in electricity prices. While by 2020 electricity prices per MWh were moving on a supply and demand-driven range between 34.67 euro/MWh at the end of the COVID-19 crisis and 75.94 euro/MWh in 2008, when there was a global banking crisis, there is now an accelerated increase amid a generalized energy deficit (Fig. 11.2).

Europe had set itself the goal of making a transition to renewable energy sources during this period, but this requires major investment in new production technologies. They were expecting higher tariffs, lower consumption and a reduction in the amount of greenhouse gases released into the atmosphere.

The European Union has taken on an important role in mitigating and combating the effects of global climate change. It has taken on five specific dimensions: energy security, decarbonization, energy efficiency, internal energy market and research, innovation and competitiveness.

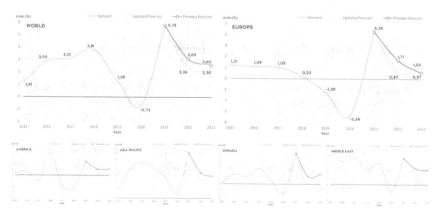

Fig. 11.1 Electricity demand developments and forecast for next year at global and regional level. *Data source* www.iea.org

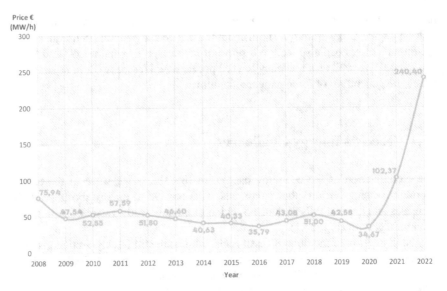

Fig. 11.2 Evolution of electricity prices at European level, 2008–2022. After data source: (www. ember-climate.org; www.energy.ec.europa.eu) *Malta and Cyprus were not taken into account for 2020–2022 period

To meet this commitment, the European Union has set energy and climate targets to be achieved by 2030 as follows:

- reducing domestic greenhouse gas emissions by at least 40% by 2030 compared to 1990;
- 32% renewable energy consumption in 2030;
- improving energy efficiency by 32.5% in 2030;
- interconnection of the electricity market to a level of 15% by 2030.

11.2   The Effects of Crises on Electricity Generation

By looking at electricity production statistics, it is easy to identify moments of crisis. At the European level, electricity production has been on an upward trend since the 1990s (2132 TWh) until the global financial crisis of 2008 (2844 TWh). With small fluctuations, a slight recovery in production followed to a value of 2800 TWh before a new downward trend in 2020 (2664 TWh) due to the COVID-19 global medical crisis (https://yearbook.enerdata.net/electricity/world-electricity-production-statistics.html).

11.2.1 Social and Political Crises

Changing political regimes have profound implications for the economy in general and the energy sector directly. In Romania, the transition from a centralized economic regime, controlled and directed according to the wishes of a small group of politicians, to a capitalist economy, with a free market and economic policies in contrast to the previous ones, has also produced a strong change in the energy sector. Total energy consumption in 1989 was 83.66 TWh, and in 1994, it reached 56.93 TWh, a drop of 32%. Consumption then steadily increased from this level to 64.7 TWh in 2008. The global economic crisis is affecting Romania's economy, so that in 2009, consumption reached only 57.6 TWh. The recovery is slow and oscillating, reaching 60.7 TWh in 2010, 56 TWh in 2020 and 59 TWh in 2021 (https://yearbook.enerdata.net/electricity/world-electricity-production-statistics.html). The decrease is also felt in the countries bordering Romania such as Serbia, Moldova or Ukraine, the trend of variation being quite similar.

Installed capacity in power generation units decreased from 21,800 MW in 1995 to 15,370 MW in 2007. Coal-fired generation units have been phased out. As a result of investments in renewable energy sources (mainly photovoltaic and wind power), the installed capacity reaches 18,309 MW in 2022. The closure of coal and gas-fired power plants has led to a fragile national energy system. At the same time, the adoption of European environmental legislation has halted investment in hydropower. Many such investments started before 1990 have remained unfinished even 30 years later. Hydropower itself is based on the renewable resource of rivers and is considered green energy. Environmental activists, however, are pushing for radical measures that impose hard-to-reach licensing or operating rules. Clearly, companies with such investments in their portfolios face difficult situations to manage. What needs to be understood is the balancing role that such investments offer to the energy system.

Electricity production from wind and solar sources has the disadvantage of a relatively short daily interval, but especially unpredictable, which makes the operation of the national and European energy system in many situations on the verge of supply interruptions.

11.2.2 Health Crisis

The most recent global health crisis is the COVID-19 crisis of 2020. Many economic sectors have suffered, and its links to the socio-economic environment are analyzed. Disruptions in various sectors, reduced mobility, stock market developments have all led to decreases in energy consumption (Jebabli 2021).

This has implicitly led to lower investment in the development of the main energy markets. The decline in energy investment (compared to 2019) was 35% in 2020 medium/downstream investment—31% in 2020, in coal supply—5% in 2020 and in fossil fuel energy—15% in 2020.

The adoption of physical distancing measures, movement restrictions and the closure of non-essential "Stay Home, Stay Safe" economic activities led to a decrease in income, which was reflected in the ability to pay energy and other utility bills (https://www.naruc.org/compilation-of-covid-19-news-resources/state-response-tracker/).

Especially during a pandemic, the survival of the most vulnerable often depends on their ability to control the temperature in their homes, use an air conditioner to relieve respiratory stress with air circulation, refrigerate medicines and food and operate medical equipment. The increased number of deaths caused by power outages since 2003 clearly demonstrates the vulnerabilities of a sudden disruption of energy services (Manes 2020).

From the perspective of energy producers, there is a decrease in energy demand during implemented home orders, ranging from 4 to 18% which represents a loss of revenue from current energy consumption (DTE 2020; Paulson 2020).

The Energy Information Administration has found in surveys that 31% of US households had problems paying their energy bills during the pandemic period, which was not the case in a normal period (Ogg 2020).

Another measure with beneficial effects in the medium and long term would be to increase investment in electricity production sources using renewable energies (wind, solar, geothermal, river and wave hydropower). Replacing closed power generation capacity as part of decarbonization measures requires large investment efforts given the much shorter operating range of renewable-based systems than fossil fuel-based systems.

11.3   Discussions

The European Union has set itself a series of long-term targets with clear intermediate milestones. It wants to achieve climate neutrality, trying to reshape European society through environmental policies.

Energy efficiency

The aim is to reduce energy end-use, so EU Member States will have to reduce their annual consumption by 1.5%. The focus will be on buildings, industry and transport. Financial support for this will be provided from the Climate Action Social Impact Mitigation Fund and revenues from emissions trading schemes (https://www.consilium.europa.eu/ro/policies/green-deal/fit-for-55-the-eu-plan-for-a-green-transition/).

Energy from renewable sources

The EU's 2030 target is to double the share of renewable energy in the energy mix. In 2020, the share was 22.10% of energy consumed (https://www.consilium.europa.eu/ro/policies/green-deal/fit-for-55-the-eu-plan-for-a-green-transition/).

Energy taxation

The energy sector is responsible for 77% of greenhouse gas (GHG) emissions in the EU, followed by agriculture with 11%, industrial processes with 9% and waste management with 3% (https://www.consilium.europa.eu/ro/policies/green-deal/fit-for-55-the-eu-plan-for-a-green-transition/).

The tax proposal concerns increasing the tax rate on the most polluting fuels (coal, oil, gas), taxing fuels used in aviation and the maritime sector. No distinction will be made between final consumers—domestic or industrial.

Review of CO_2 emission limits for new cars

Cars and vans are responsible for 15% of all CO_2 emissions. The aim is to have zero emissions by 2030. The current limit (2021–2024) is 147 g/km. It is estimated that between 900 and 1100 Mtoe (diesel and petrol) will be saved between 2030 and 2050 (https://www.consilium.europa.eu/ro/policies/green-deal/fit-for-55-the-eu-plan-for-a-green-transition/). It should not be forgotten that cars and vans currently produced have a service life that will exceed 2030. In these circumstances, the fall in demand for diesel and petrol will also lead to a fall in their price, which will indirectly stimulate and make some transport more efficient.

Improving infrastructure for the use of alternative fuels

Increase the number of electric charging stations for cars and vans, hydrogen or liquefied natural gas filling stations. Such facilities will also be extended to aircraft stationed on the ground as well as ships in ports.

Relocation of CO_2 emissions.

Industries with high CO_2 emissions through differentiated taxation could be forced to relocate to other countries with more permissive environmental policies.

CO_2 storage at biomass level

Europe's forests absorb 10% of total gas emissions every year. The aim is to balance the balance between emissions from human activities and the amounts absorbed and stored by forests. The 2030 target is to reach 310 Mt CO_2 equivalent absorbed (https://www.consilium.europa.eu/ro/policies/green-deal/fit-for-55-the-eu-plan-for-a-green-transition/).

These targets proposed by the EU for 2030 and 2050 will be achieved at great sacrifice by the population and investors within the union. To this end, a fund is to be set up to mitigate the social impact of climate action. The fund will be made up of the sale of certificates at the carbon price for emissions from economic activities or fuels. The beneficiaries will be vulnerable households, an estimated 34 million people, microenterprises and transporters. The budget allocated as funding to Member States is €59bn for the period 2027–2032 (https://www.consilium.europa.eu/ro/policies/green-deal/fit-for-55-the-eu-plan-for-a-green-transition/).

The targets are obviously ambitious and necessary in the current context of climate change. However, the implementation of the program implies its aggregation with other parameters that I believe have not been carefully evaluated. The emergence of unexpected crises such as the global health crisis of COVID-19, or the regional crisis generated by the armed conflict between Russia and Ukraine, may divert the chosen route. The different levels of development of member countries, the income of their citizens, but above all the willingness of the population to make certain efforts and sacrifices from their current comfort may be a cause of deviation from the schedule.

The developers of the decarbonization and climate neutrality plan do not seem to understand that the earth has a single atmosphere, which is not divided into continents (Regulation (EU) 2018). This gaseous envelope exhibits legacies and processes that occur on a regional and global scale. Climate is the totality of meteorological processes and phenomena that characterize the average state of the atmosphere of a region. For example, El Niño occurs in the tropical, Southern Pacific Ocean due to the warming of ocean surface water and its movement from the west to the east (Western South America) generating droughts, storms and floods. This phenomenon generates climate anomalies across the planet.

In other words, the developers of these concepts and measures do not see the phenomenon on its true scale. It is as if there are seven people in a room who smoke and one of them gives up the habit. This does not mean that after quitting smoking, they will breathe clean air.

Stimulating the migration of polluting industries from Europe to other continents or countries (which have not joined the Kyoto Protocol) does nothing to solve the problem of air pollution but creates a false idea of solving it (UNFCCC 1997; Regulamentul (UE) 2021). Europe still needs aluminum, steel, cement, chemical fertilizers or electricity. Even if they are not produced on the European continent, their production will still contribute to global air pollution. In the cement industry, China is the largest producer, followed by India and then the United States. Europe produces only about 10% of the world's cement (Vanderborght and Brodmann 2001). But even on the European continent, there is a trend toward the migration of production units to developing countries. In Romania, there are 7 plants, of which 3 are German-owned and two are Swiss or Irish-owned.

Perhaps, it would be more appropriate to invest in research and technologies to clean up industry and transport respectively so as to keep activity in European locations, with all that is involved in the horizontal economic branches, so that they become emission neutral.

Continental climate neutrality does not exist. In terms of zero impact on greenhouse gas emissions, it is necessary to explain that we will continue to be exposed to the effects of global climate change (https://ec.europa.eu/info/law/better-regulation/have-your-say/initiatives/12227-Revision-of-the-Energy-Tax-Directive-/public-consultation_en). Increases in average air temperature, heat waves, violent winds, droughts, desertification, floods, rising global ocean levels will continue to occur on the European continent (Diaconu 2018).

Producing zero CO_2 emission cars means electrifying the car fleet. The rest of the non-fossil fuel technologies are still in the early stages of development. Reaching this target does not mean that CO_2 emissions in the transport sector have been solved. No data are presented on the carbon footprint of electric car battery production, battery life and especially battery recycling. Freight transport will still be a major consumer of fossil fuels, given the long operating time of ships, planes and trucks.

With the relocation of polluting production facilities to neighboring countries or other continents, the transport of goods (raw materials or finished products) will increase, generating new energy consumption.

At the same time, as long as at least half of the primary energy sources with greenhouse gas emissions are in the electricity production basket, we can only speak of a relocation of pollution from congested urban areas to the outside. We will have cleaner air in cities where people have incomes that allow them to buy more expensive cars than those with internal combustion engines. Instead, we will see an increase in deteriorating health in low-income areas and countries where the medical system is poor. The aim is to reduce medical expenditure and costs, but with a transfer of mortality and lower life expectancy to the poor.

These measures adopted or proposed by Europe also indicate the technological inability to find technical solutions to reduce the planet's pollution.

11.4 Conclusions

Energy is linked to key areas such as water supply, health, food production, transport and will continue to be the main driver of the global economy in the future. The social trend of global development will require ever greater quantities of energy; thus, the risk of running out is possible.

The exploitation of primary energy resources with greenhouse gas emissions and low energy yields will have to be removed from the energy production mix. This will reduce dependence on coal, oil and gas resources, so that their price variations will no longer cause imbalances in fragile economies. The rise or fall of nations, with its implications, will no longer be driven by access to energy resources, and the struggle for them will no longer generate conflict.

Identifying new energy sources, new energy production and transport technologies will ensure future socio-economic development. This will limit the number of armed conflicts and acts of terrorism, environmental pollution and loss of life.

Solutions to the energy crisis are now expected to come from the exploitation of renewable energy sources. Thus, increased action is expected in the following directions:

• Developing electricity generation capacity from renewable sources (hydro, wind, solar, biomass). Many of these areas shift pressure and dependence from fossil energy resources to other components of the environment (water resources, biomass, vegetation, soil).

Advantages	Disadvantages
Rapid possibility that administrative measures can reduce energy consumption (reduction of ambient temperature inside buildings, public lighting, etc.)	High commissioning time for such investments (dams, wind fields, tidal power plants)
	High investment costs, dependence on existing transmission grids
Opportunities	Relatively short production intervals compared to fossil fuel or nuclear power plants
Use of degraded land	High variability of production time intervals
Decreasing the rate of desertification	Occupation of large areas of land (in the case of PV)
Possibility of energy independence without the necessity of fossil fuels	Limited production and installation capacities of energy systems in the short term
	Need to train personnel with technical skills in these new areas

- Energy efficiency: reducing fossil fuel consumption and overall energy consumption, by making buildings more thermally efficient, changing consumption habits, reducing waste, increasing the performance of transport networks and especially storage.

Advantages	Disadvantages
Systemic crises have accelerated the transition to renewable energy sources	Inertia of the population to give up current comfort conditions
A joint approach to measures makes the transition to a renewable energy system easier	Structural measures to make buildings more efficient involve high costs in the short term

- Developing the necessary infrastructure for energy production, transmission and storage. This means that energy must have the necessary infrastructure to circulate and cover all regions. Interconnection plans should be designed to reduce emerging supply risks.

Advantages	Disadvantages
Low greenhouse gas emissions	Huge costs for energy storage in batteries
Low energy production costs	Different energy policies at continental level
Possibility of developing sources with small production capacities	Design of new transmission networks on geographically or politically difficult routes
Training small investors in financing new production units	Underestimated environmental costs

References

Azzuni A, Breyer C (2018) Definitions and dimensions of energy security: a literature review. WIRES Energy Environ 7:e268

Boin A, Hart P, Stern E, Sundelius B (2005) The politics of crisis management: public leadership under pressure. Cambridge University Press, Cambridge, UK, New York

Congleton RD (2005) The political economy of crisis management: surprise, urgency, and mistakes in political decision making. In: Kurrild-Klitgaard P (ed), The dynamics of intervention: regulation and redistribution in a mixed economy, Elsevier, Amsterdam

Coyne CJ (2011) Constitutions and crisis. J Econ Behav Organ 80:351–357

Diaconu DC (2018) Water from a geographic perspective. Transversal Publishing House, Târgoviște, 161 p. ISBN: 978-606-605-183-5 (in Romanian)

DTE (2020) DTE energy launches new personalized service protection program to help customers impacted by COVID-19 protect their energy service

Dyduch J, Skorek A (2020) Go South! Southern dimension of the V4 states' energy policy strategies—an assessment of viability and prospects. Energy Policy 140:111372

Grossman PZ (2013) U.S. energy policy and the pursuit of failure, Cambridge University Press, Cambridge

https://ec.europa.eu/eurostat/statistics-explained/index.php?title=Electricity_production,_consumption_and_market_overview. Accessed in 10 Nov 2022

https://ec.europa.eu/info/law/better-regulation/have-your-say/initiatives/12227-Revision-of-the-Energy-Tax-Directive-/public-consultation_en

https://yearbook.enerdata.net/electricity/world-electricity-production-statistics.html. Accessed in 11 Nov 2022

https://www.consilium.europa.eu/ro/policies/green-deal/fit-for-55-the-eu-plan-for-a-green-transition/

https://www.naruc.org/compilation-of-covid-19-news-resources/state-response-tracker/. Accessed in 11 Nov 2022

https://www.statista.com/statistics/302233/china-power-generation-by-source/. Accessed in 10 Nov 2022

Jebabli I (2021). Financ Res Lett. https://doi.org/10.1016/j.frl.2021.102363

Keeler JTS (1993) Opening the window for reform: mandates, crises and extraordinary policy-making. Comp Polit Stud 25:433–486

Manes N (2020) Consumers, DTE show declines in business energy use and rising home use amid covid-19. April 28

Nukusheva A, Ilyassova G, Rustembekova D, Zhamiyeva R, Arenova L (2020) Global warming problem faced by the international community: international legal aspect. Int Environ Agreem 1:1–15

Nyga-Łukaszewska H, Aruga K, Stala-Szlugaj K (2020) Energy security of poland and coal supply: price analysis. Sustainability 12:2541

Ogg JC (2020) 15 basic economy dividend stocks you'll want to own after the coronavirus recession

Paulson LD (2020) COVID-19 impacts on energy demand, infrastructure yet to be known

Regulamentul (UE) 2021/1119 al Parlamentului European și al Consiliului din 30 iunie 2021 de instituire a cadrului pentru realizarea neutralității climatice și de modificare a Regulamentelor (CE) nr. 401/2009 și (UE) 2018/1999 („Legea europeană a climei") (JO L 243, 9.7.2021, p 1)

Regulation (EU) 2018/842 of the European Parliament and of the Council of 30 May 2018 on the annual mandatory reduction of greenhouse gas emissions by Member States in the period 2021–2030 to contribute to climate action to meet their commitments under the Paris Agreement and amending Regulation (EU) No 525/2013 (OJ L 156, 19.6.2018, p 26)

Salonen H (2018) Public justification analysis of Russian renewable energy strategies. Polar Geogr 41:75–86

UNFCCC (1997) Kyoto protocol to the United Nations framework convention on climate change adopted at COP3 in Kyoto, Japan, on 11 Dec 1997

Vanderborght B, Brodmann U (2001) The cement CO_2 protocol: CO_2 emissions monitoring and
    reporting protocol for the cement industry
www.ember-climate.org. Accessed in 10 Oct 2022
www.energy.ec.europa.eu. Accessed in 11 Nov 2022
www.iea.org. Accessed in 10 Nov 2022

Printed in the United States
by Baker & Taylor Publisher Services